AF595110

Titanium and Its Alloys for Biomedical Applications

Titanium and Its Alloys for Biomedical Applications

Editor

Hyun-Do Jung

MDPI • Basel • Beijing • Wuhan • Barcelona • Belgrade • Manchester • Tokyo • Cluj • Tianjin

Editor
Hyun-Do Jung
Catholic University of Korea
Korea

Editorial Office
MDPI
St. Alban-Anlage 66
4052 Basel, Switzerland

This is a reprint of articles from the Special Issue published online in the open access journal *Metals* (ISSN 2075-4701) (available at: https://www.mdpi.com/journal/metals/special_issues/Titanium_Biomedical_Applications).

For citation purposes, cite each article independently as indicated on the article page online and as indicated below:

LastName, A.A.; LastName, B.B.; LastName, C.C. Article Title. *Journal Name* **Year**, *Volume Number*, Page Range.

ISBN 978-3-0365-4935-4 (Hbk)
ISBN 978-3-0365-4936-1 (PDF)

Cover image courtesy of Hyun-Do Jung.

Contents

About the Editor

Hyun-Do Jung

Dr. Hyun-Do Jung, an assistant professor, is currently working at the Department of BioMedical-Chemical Engineering (BMCE), Catholic University of Korea, Korea. He is the Principal Investigator of the Advanced Biomaterials for Biomedical Applications Laboratory (ABBA Lab). His laboratory is trying to achieve high efficiency and performance in biomedical applications. He is particularly interested in developing advanced biomaterial systems to address limitations in hard (e.g., dental and orthopedic) and soft (e.g., cardiovascular, skin, and artificial organ) tissue engineering, with the general approach of creating a scaffold with an optimal design and modifying its surface to improve its performance in the physiological environment. He received his PhD in Materials Science and Engineering from Seoul National University, Korea. With more than a decade of experience in academia and industry, he has published more than 70 papers in peer-reviewed journals and has more than 20 registered patents. He is a committee member of the Korean Institute of Metals and Materials (KIMM), the Korean Tissue Engineering and Regenerative Medicine International Society (KTERMS), and the Korean Society for Biomaterials (KSB).

 *metals*

Editorial

Titanium and Its Alloys for Biomedical Applications

Hyun-Do Jung [1,2]

[1] Department of Biomedical-Chemical Engineering, Catholic University of Korea, Bucheon 14662, Korea; hdjung@catholic.ac.kr

[2] Department of Biotechnology, The Catholic University of Korea, Bucheon 14662, Korea

Citation: Jung, H.-D. Titanium and Its Alloys for Biomedical Applications. *Metals* **2021**, *11*, 1945. https://doi.org/10.3390/met11121945

Received: 21 November 2021
Accepted: 26 November 2021
Published: 2 December 2021

1. Introduction and Scope

In the past decades, metals have been considered as promising materials in the fields of regenerative medicine and tissue engineering. Metallic bio-materials with excellent mechanical strength can effectively support and replace damaged tissue. Hence, metals have been widely used in load-bearing applications for dentistry and orthopedics. cobalt-, iron-, and titanium (Ti)-based alloys are representative bio-metals used in various forms such as vascular stents, hip joints, dental, and orthopedic implants. However, the alloying elements of Co- and Fe-based alloys, Co, Ni, and Cr, induce severe toxicity when ionized in the body, which limits their clinical use.

On the other hand, Ti and its alloys have been widely used as medical devices and implants applied in dental and orthopedic applications due to their excellent bone regeneration ability, mechanical properties, and corrosion resistance. Even though Ti and its alloys have been generally used for biomedical applications, there are still challenges that must be met in order to satisfy clinical applications. For example, osseointegration with the surrounding bone tissue at the initial stage of implantation has been pointed out as a major issue. In recent years, much attention has been placed on the design of new Ti alloys and composites with a modified process such as the advanced casting method and artificial intelligence for fabricating dental and orthopedic implants with lower elastic moduli in combination with higher strength. On the other hand, some researchers have focused on the surface modification of Ti-based medical devices and implants for inducing rapid bone ingrowth.

This Special Issue, "Titanium and Its Alloys for Biomedical Applications", has been proposed as a means to present recent developments in biomedical applications. The nine research articles included in this Special Issue cover broad aspects of Ti-based alloys and composites with respect to their composition, mechanical, and biological properties, as highlighted in this editorial.

2. Contributions

2.1. Surface Modification

Four papers in this section concern the surface modification of Ti and its alloys, including the final developmental stage of medical devices and surface treatment of implants.

Kim et al. studied the ultraprecision magnetic abrasive finishing (UPMAF) of a nickel (Ni)–Ti stent wire using ecofriendly oil [1]. Generally, the accuracy and precision of Ni–Ti stent wire sizes are important factors in the clinical field as well as toxicity of the stent surface. Therefore, in this study, a UPMAF process with earth-friendly olive and castor oil was investigated. Using both olive and castor oil, the surface roughness was modified without any change of components of original Ni–Ti metal compared to using industrial light oil.

Kim et al. compared and analyzed various Ti surface treatment technologies for the orthopedic field [2]. They applied hydroxyapatite (HA) blasting, sand blasting and an acid-etching (SLA) process on a Ti substrate. Additionally, other anodic oxidation (AO)

and micro-arc oxidation (MAO) processes were conducted on the SLA-treated Ti. A TiO_2 layer created by the AO and MAO processes increased porosity and surface roughness and consequently enhanced cellular responses.

Huang et al. used high-power impulse magnetron sputtering to deposit tantalum oxide and Ta (Zn)O thin films on porous Ti prepared by plasma electrolytic oxidation [3]. The surface showed antibacterial properties with Staphylococcus aureus and Actinobacillus actinomycetemcomitans bacterias and improved cellular activities in terms of the adhesion, migration, and proliferation of osteoblastic cells.

Oh et al. used electrophoretic deposition to modify the Ti surface and layered graphene oxide-based coatings that control drug loading and release by varying the thickness of the coating layer [4]. The coatings enhanced surface hydrophilicity and hardness as well as bone differentiation properties when compared to the drug simply adsorbed to the surface.

2.2. Alloys and Ti Matrix Composite

The Ti alloys and Ti matrix composite were developed for use in the aerospace industries by enhancing mechanical properties and corrosion resistance. Despite this, it was revealed that such alloys can also find an excellent application field as biomaterials. Therefore, this section comprises three papers and concerns the manufacturing techniques of alloy and composite applied in bioengineering.

Jörg et al. fabricated a Ti6Al4V alloy that is extensively used in biomedical applications, using precision centrifugal casting [5]. The optimized centrifugal casting process of the Ti6Al4V provided a high-quality surface and equivalent biologic material quality as well as a reduction in cost for manufacturing.

Jha et al. used the CALPHAD and artificial intelligence for new Ti-based alloys [6]. They proposed an efficient approach for screening the compositions and process parameters that maximize the stability of β phase while minimizing α'' and ω-phase formations on Ti–Zr–Nb–Sn systems. In order to accelerate the development process, AI algorithms were used to predict alloys that are expected to meet the requirements with the phase stability of β phase.

Ti-matrix composites using bone were produced by Jeong et al. [7]. Ball-milled Ti6Al4V powders and equine bone (EB) powders were spark plasma sintered at 1000 °C. The existence of hydroxyapatite, the main component of EB, in a sintered Ti6Al4V matrix increased the hardness of composites, indicating good mechanical properties.

2.3. Advanced Analysis

The characterization and evaluation of Ti alloys have also attracted significant attention to keep up with development speed of Ti-manufacturing technologies. Two papers discussed advanced analysis methods for mechanical and biological properties

There have been many efforts to fabricate Ti alloys with a low elastic modulus to reduce stress-shielding effects that deteriorate the bone healing. Meng et al. investigated the deformation behavior of a Ti36Nb5Zr alloy with a β phase using in situ synchrotron X-ray diffraction (SXRD) [8]. They proved deformation mechanisms of β-type Ti alloys including elastic and plastic deformations.

Beltrán et al. confirmed the potential of grade V Ti mini-transitional implants for a temporary overdenture [9]. Bone area fraction occupancy after in vivo test was analyzed by backscattered scanning electron microscopy (BS-SEM), and eventually it was demonstrated that grade V Ti is an appropriate material for a dental implant.

3. Conclusions and Outlook

The present Special Issue of *Metals* presents an extensive insight into the state-of the-art of the research on developing Ti and its alloys for biomedical applications. The number of articles and the wide range of topics prove the continued interest in this group of materials. The development and progress in Ti and its alloys have played an important role in manufacturing medical devices and dental and orthopedic implants that are not possible with other metals.

As a guest editor of this Special Issue, I hope that these papers will be useful to researchers, designers, and engineers investigating in the development of advanced Ti alloys and improvement of their performance. Additionally, I am honored to be a guest editor and glad to report the success of this Special Issue. My sincere thanks to the reviewers and editors of *Metals* for their continuous support.

Funding: This work was supported by The Catholic University of Korea, Research Fund, 2021 and the National Research Foundation of Korea (NRF) grant funded by the Korea government (MSIT) (2021R1A2C1091301), the framework of international cooperation program managed by the National Research Foundation of Korea (2021K2A9A2A06037540), and Korean Fund for Regenerative Medicine funded by Ministry of Science and ICT, and Ministry of Health and Welfare (2021M3E5E5096420, Korea).

Conflicts of Interest: The author declare no conflict of interest.

References

1. Kim, J.S.; Nam, S.S.; Heng, L.; Kim, B.S.; Mun, S.D. Effect of environmentally friendly oil on Ni-Ti stent wire using ultraprecision magnetic abrasive finishing. *Metals* **2020**, *10*, 1309. [CrossRef]
2. Kim, J.; Lee, H.; Jang, T.-S.; Kim, D.; Yoon, C.-B.; Han, G.; Kim, H.-E.; Jung, H.-D. Characterization of Titanium Surface Modification Strategies for Osseointegration Enhancement. *Metals* **2021**, *11*, 618. [CrossRef]
3. Huang, H.-L.; Tsai, M.-T.; Chang, Y.-Y.; Lin, Y.-J.; Hsu, J.-T. Fabrication of a Novel Ta (Zn) O Thin Film on Titanium by Magnetron Sputtering and Plasma Electrolytic Oxidation for Cell Biocompatibilities and Antibacterial Applications. *Metals* **2020**, *10*, 649. [CrossRef]
4. Oh, J.-S.; Jang, J.-H.; Lee, E.-J. Electrophoretic Deposition of a Hybrid Graphene Oxide/Biomolecule Coating Facilitating Controllable Drug Loading and Release. *Metals* **2021**, *11*, 899. [CrossRef]
5. Jörg, F.; Betül, K.A.; Heiner, M.; Gerhard, K.; Erwin, B.R. Effects of Ti6Al4V Surfaces Manufactured through Precision Centrifugal Casting and Modified by Calcium and Phosphorus Ion Implantation on Human Osteoblasts. *Metals* **2020**, *10*, 1681. [CrossRef]
6. Jha, R.; Dulikravich, G.S. Discovery of new Ti-based alloys aimed at avoiding/minimizing formation of α'' and ω-phase using CALPHAD and artificial intelligence. *Metals* **2021**, *11*, 15. [CrossRef]
7. Jeong, W.; Shin, S.-E.; Choi, H. Microstructure and Mechanical Properties of Titanium–Equine Bone Biocomposites. *Metals* **2020**, *10*, 581. [CrossRef]
8. Meng, Q.; Li, H.; Wang, K.; Guo, S.; Wei, F.; Qi, J.; Sui, Y.; Shen, B.; Zhao, X. In situ synchrotron X-ray diffraction investigations of the nonlinear deformation behavior of a low modulus β-Type Ti36Nb5Zr alloy. *Metals* **2020**, *10*, 1619. [CrossRef]
9. Beltrán, V.; Weber, B.; Lillo, R.; Manzanares, M.-C.; Sanzana, C.; Fuentes, N.; Acuña-Mardones, P.; Valdivia-Gandur, I. Histomorphometric Analysis of Osseointegrated Grade V Titanium Mini Transitional Implants in Edentulous Mandible by Backscattered Scanning Electron Microscopy (BS-SEM). *Metals* **2021**, *11*, 2. [CrossRef]

 *metals*

MDPI

Article

Electrophoretic Deposition of a Hybrid Graphene Oxide/Biomolecule Coating Facilitating Controllable Drug Loading and Release

Jun-Sung Oh, Jun-Hwee Jang and Eun-Jung Lee *

Department of Nano-Biomedical Science & BK21 PLUS NBM Global Research Center for Regenerative Medicine, Dankook University, Cheonan 31116, Korea; gda4101@dankook.ac.kr (J.-S.O.); junhweej@dankook.ac.kr (J.-H.J.)
* Correspondence: leeej@dankook.ac.kr; Tel.: +82-41-550-3697

Abstract: Two-dimensional (2D) graphene oxide (GO) exhibits a high drug loading capacity per unit mass due to its unique structure and hydrophilicity and has been widely researched for drug-delivery systems. Here, we modified the surfaces of metal implants; we applied GO-based coatings that controlled drug loading and release. We used electrophoretic deposition (EPD) to apply the coatings at room temperature. The EPD coatings were analyzed in terms of their components, physical properties such as hardness and hydrophilicity, and in vitro cell tests of their biological properties. Uniform GO-EPD coatings improved surface hydrophilicity and hardness and greatly improved the bone differentiation properties of the metal substrate. Drug loading and release increased greatly compared to when the drug was adsorbed to only the surface of a coating. GO facilitated deposition of a drug-containing coating via EPD, and the surface modification, and drug loading and release, were controlled by the thickness of the coating.

Keywords: graphene oxide; electrophoretic deposition; implant; biomolecule; complex

Citation: Oh, J.-S.; Jang, J.-H.; Lee, E.-J. Electrophoretic Deposition of a Hybrid Graphene Oxide/Biomolecule Coating Facilitating Controllable Drug Loading and Release. *Metals* **2021**, *11*, 899. https://doi.org/10.3390/met11060899

Academic Editor: Hyun-Do Jung

Received: 16 April 2021
Accepted: 28 May 2021
Published: 31 May 2021

1. Introduction

Metals are widely used to replace damaged bones, especially load-bearing bones [1]. Any metal is a bioinert material, the use of which raises concerns about (poor) biocompatibility, inappropriate mechanical properties, and inflammatory/immune reactions caused by metal ion dissolution [2]. Surface modifications of dental or orthopedic metal-based implants enhance biocompatibility and functionality [3,4]. Surface treatments may be physical, chemical, or biological in nature. Physical methods may alter the surface morphology to induce attachment to regenerated bone, or oxidize the implant surfaces to increase hydrophilicity and reduce corrosion caused by micro-arc oxidation and anodizing [5]. Chemical methods alter the surface of implants without significantly affecting their bulk properties, yielding hard, wear-resistant hydrophilic surfaces. The various techniques include chemical vapor deposition (CVD), plasma vapor deposition (PVD), ion-beam deposition (IBAD), grafting techniques, and the use of self-assembling monolayers (SAMs) [6–9]. Biological methods effectively improve the biological properties of bioinert metal implants [10].

Synthetic biomimetic strategies enhancing the functionality of metal-based implants have focused principally on the addition of biomolecules to implant surfaces. Growth factors and protein-mimetic peptides improve the interactions between the implant and the biological environment, with preservation of the bulk implant's mechanical properties [11]. Reactive groups are required for biomolecular tethering. However, bioinert metal surfaces lack such groups. Surface active groups (e.g., –OH, –COOH, and –NH2) are essential for surface modification. Oxygen-terminal carbon-based materials facilitate strong physisorption of biomolecules to carbon-based materials [12].

Graphene oxide (GO) contains several reactive oxygen groups (e.g., C=O, COOH, OH, and C-O-C), suspends well in water, and interacts with biomolecules and drugs. The unique flake-type two-dimensional (2D) structure is associated with high drug loading per unit of GO mass; the GO surface area is high. GO exhibits good mechanical properties, good biocompatibility (especially in terms of osteoconductivity), and good antimicrobial activity. GO is an optimal coating for orthopedic implants [13,14]. There are many reports that carbon-based materials, such as graphene, combined with biomolecules (BM) are effective in regenerating bone tissue. Examples include complexes of carbon-based materials with BM such as BMP-2, FGFs, and Simvastatin [15–17]. In addition, these complexes have been produced in various forms, including film, coating, particles, scaffolds, and fibers, and the properties analyzed and widely applied in implants research for tissue regeneration [18–21]. Biomimetic surface modifications enhance implant function; however, biomolecules are vulnerable to high temperatures, strong acid/base conditions, and chemical solvents [22]. Therefore, many studies use natural biopolymers as a base material that can be processed in aqueous conditions to prevent the stability of biomolecules, or incorporate biomolecules after the fabricating process of base materials is completed [21,23]. In particular, the surface modifications of metal implants require the gentle condition for all processes including the preparation of BM-combined composites and deposition of them on the surface. The electrophoretic deposition (EPD) method can be used to form coatings from aqueous solutions at room temperature. EPD deposits colloidal particles in an aqueous electrolytic bath onto substrates. The coating time is short, and the coatings are uniform and continuous [24,25]. EPD has been used to produce graphene films, graphene-based reinforced composites, complex materials, interleaved porous structures, and nanoparticle-spaced graphene films [26]. To modify the surfaces of implants used for hard-tissue engineering, researchers have sought to reduce internal corrosion, increase hardness, and enhance biocompatibility by the addition of biopolymers; however, few studies have explored combinations of GO with therapeutic drugs. We are the first to use EPD to develop GO-biomolecule (GO-BM) hybrid coatings of controllable thickness; the coatings contain large amounts of drugs. If BMs are exposed on an implant surface, an additional layer is required to protect the BMs from loss or denaturation during transplantation. Our method reduces BM damage and allows control of drug loading and release. It is a technology applicable to drug-eluting stents or orthopedic implants development, which is expected to lead to enhanced therapeutic effects. In this study, GO-EPD coatings for biomedical applications were evaluated in terms of composition, physical properties, cellular interactions, and drug release [27–30].

2. Materials and Methods

2.1. Materials

Titanium plates (bare Ti of grade 2, Titanart, Incheon, Korea) served as EPD substrates. The plates were polished with 800- and 1200-grit silicon carbide paper and ultrasonically cleaned in acetone, ethanol (Duksan, Ansan, Korea), and distilled water (5 min for each bath). Single-layer GO was from Graphene Supermarket (Ronkonkoma, NY, USA). Ethanol (Duksan, Ansan, Korea) was 99.6% pure. Green fluorescence protein (GFP) and bone morphogenic protein-2 (BMP-2) were supplied by Genoss, Gyeonggi-do, Korea.

2.2. Electrophoretic GO Deposition

A GO suspension (500 μg/mL) was prepared in 80% (v/v) ethanol and sonicated for 15 min. Ti plates were submerged in the suspension at 1 cm from the cathode, and EPD proceeded at 10 mA for 1 min. The plates were then dried and stored in a desiccator. Prior to biological tests, the coated plates were sterilized with 70% (v/v) ethanol. EPD was also used to prepare drug-eluting coatings; the test BMs were GFP and BMP-2. Various amounts of the BMs (10, 25, 50, and 100 μg/mL) were added to GO suspensions or GO-coated Ti plates, magnetically stirred for 3 h, and EPD was performed or drug release was assessed.

2.3. Characterization

The surface morphology and coating thicknesses were studied via scanning electron microscopy (SEM; JSM-6510, JEOL Ltd., Tokyo, Japan). X-ray diffraction (XRD; Ulima IV, Rigaku, Tokyo, Japan) was used to define the phase compositions of Ti and GO-coated Ti. The test voltage was 40 kV and the current was 40 mV; Cu-Kα radiation was delivered (λ = 1.540598 Å) over a 2θ range of 5–70° with a step size of 1° and a count time of 1 min/step. A Raman spectroscopy (DXR2xi, Thermo Fisher, Waltham, MA, USA) was performed at 532 nm. Zeta potentials were measured using a Zetasizer (Nano-ZS, Malvern Instruments, Malvern, UK) with water as the dispersant. The electrophoretic mobilities of suspensions were converted to zeta potentials. An X-ray photoelectron spectroscopy (XPS; Axis Supra, Kratos, UK) was performed with the aid of focused, monochromatized Al Ka radiation (hν = 1486.6 eV). An atomic force microscope (AFM) (Bruker Dimension Edge, Middlesex County, MA, USA) was used to characterize surface microstructure and morphology. Coating hardness was measured using a Vickers indenter (HM-221, Mitutoyo, Kanagawa, Japan) at a load of 0.98 N. Contact angles (D7334-08 device, ASTM, Montgomery County, PA, USA) were used to measure the surface wettabilities of Ti and GO-coated Ti plates.

2.4. BM Loading and Release

BM (100 µg/mL) loading and release into/from GO before and after GO-EPD were evaluated by visualizing the GFP via confocal laser scanning spectroscopy (CLSM, Zeiss-LSM510, Carl Zeiss, Oberkochen, Germany). The extent of fluorescence reflected the BM level. BM loaded onto and then released from GO-BM/Ti coatings was measured by ELISA after immersing the complexes in phosphate-buffered saline (PBS) pH 7.4 at 37 °C for up to 20 days. The PBS was replaced at defined intervals. The released BM levels were measured by deriving optical absorbances at 490 nm using a microplate reader (SpectraMax M series, Molecular Devices, San Jose, CA, USA).

2.5. Cell Morphology

For the in vitro cell tests, mesenchymal stem (mMSC) cells were isolated from mouse (5 weeks, male) bone marrow harvested from the tibia and femoral marrow compartments, then cultured in general cell media, utilizing Dulbecco's Modified Eagle's Medium (DMEM, Welgene, Gyeongsan, Korea) supplemented with 10% fetal bovine serum (FBS, Gibco, Eri County, NY, USA), and 1% penicillin/streptomycin (P/S, ThermoFisher, Waltham, MA, USA) at 37 °C, with 5% CO_2, and at 90% humidity. Cells were seeded onto Ti and GO-coated Ti plates at 1×10^4 cells/mL. After being cultured for 4 h or 3 days, cells were fixed in 4% (*v/v*) paraformaldehyde (Sigma-Aldrich, St. Louis, MO, USA) with Triton X-100 for 10 min and rinsed three times in PBS. Then, 200 µL of Alexa 647 (red) and 488 (green) solutions (Thermo Fisher Scientific, Waltham, MA, USA) were added to each well followed by incubation for 1 h. The stained cells were rinsed three times with PBS and observed under a confocal laser scanning microscope.

2.6. Alkaline Phosphatase (ALP) Activity

Alkaline phosphatase (ALP) activity was measured after 7 days of culture using the para-nitrophenyl phosphate assay (Takara, Tokyo, Japan) according to the manufacturer's protocol. Cells were washed in DPBS and lysed in 0.1% (*v/v*) Triton X-100. Proteins in the extracts were quantified using a BCA protein assay kit (Takara, Tokyo, Japan). The absorbance of the reaction product was measured at 405 nm. ALP activity was normalized to the total protein content.

2.7. Alizarin Red S Staining to Detect Mineralization

Alizarin Red S (ARS) staining was used to detect extracellular calcium deposits generated by 14 days. Cells were washed twice in DPBS, fixed in 4% (*v/v*) paraformaldehyde for 15 min, and stained with 2% (*w/v*) ARS solution (pH 4.2). The cells were then washed three times with distilled water and dried at room temperature. Mineralization-positive cells were stained red. To quantify staining, the stain was extracted into 10% (v/v) acetic acid for 30 min, followed by neutralization (ammonium hydroxide), and the absorbances were read at 405 nm.

2.8. Statistical Analysis

The quantitative results of the in vitro cell tests were collected in at least three replicates from each test group. The statistical analyses were performed using a t-test, and comparisons between groups were analyzed by a one-way analysis of variance test. The differences with a p value < 0.05 were considered statistically significant (* $p < 0.05$).

3. Results and Discussion

3.1. Preparation of GO-Coated Ti Plates

GO-coated Ti plates were prepared via EPD at room temperature. A schematic is shown in Figure 1a. Hydrophilic GO bound BMs between the many GO layers (Figure 1b) [31]. Prior to EPD, BM was conjugated onto GO sheets and the complexes were evenly dispersed in electrolytic baths with 80% (v/v) ethanol; EPD followed. The GO coatings thus contained internal BM. The schematic of Figure 1b shows how BM was attached after the GO coating. The GO coating thickness can be controlled when modifying metal implants; the coating can contain large amounts of BM. If the BM were to be exclusively surface-attached, a protective layer would be required. Our method removed the need for such a layer [32].

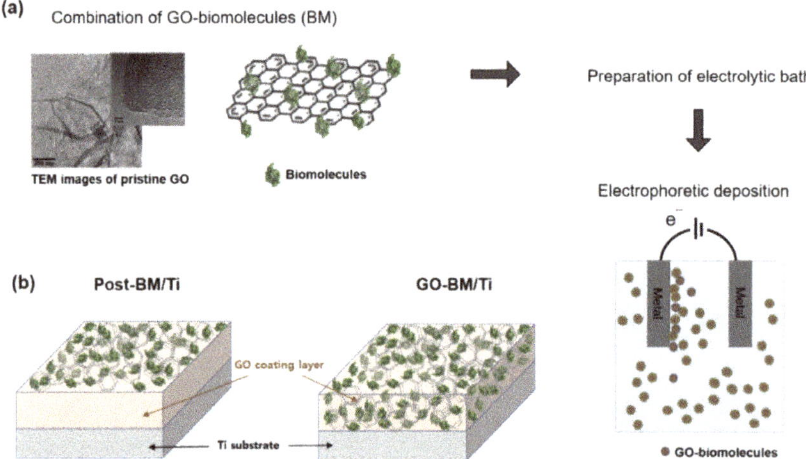

Figure 1. Schematic diagram of electrophoretic deposition (**a**). The coating layers termed Post-BM/Ti and GO-BM/Ti (**b**).

3.2. Characterization of GO-Coated Ti Plates

GO coating morphology and thickness depend on the EPD time, voltage, current, and GO concentration. Figure 2a shows photographs of bare Ti and GO-coated Ti plates. After EPD, the Ti substrate was uniformly covered with brown GO. Figure 2b,c,e shows SEM images of bare and GO-coated Ti plates; a short deposition time created thin films and a long deposition time created thick films.

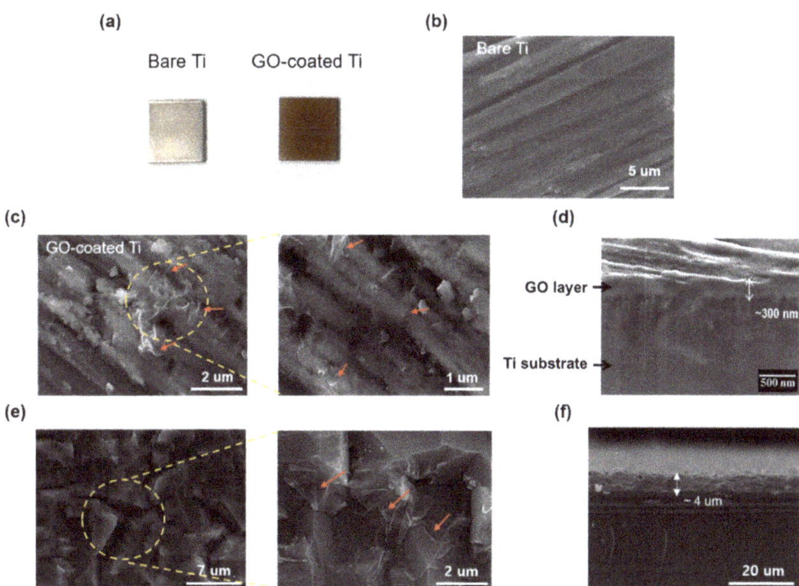

Figure 2. Optical surface images (**a**). SEM image of a bare Ti surface (**b**). Images of Ti plates coated thinly and thickly with GO (the arrows indicate GO flakes) (**c**,**e**). Cross-sectional views of thin and thick GO coatings (**d**,**f**).

Figure 2c,e shows the morphologies of (smooth) bare and GO-coated Ti plates. Cross-sections were prepared to measure coating thickness by EPD time. Increasing the time from 30 to 600 s increased the coating thickness. Figure 2d shows that the 30-s layer was less than 300 nm thick; Figure 2f shows that the 10-min thickness was approximately 4 µm.

The coated GO layer was analyzed by X-ray photon and Raman spectroscopy, and an X-ray diffraction analysis. The XPS spectra of GO-coated and bare Ti revealed titanium (Ti2p), oxygen (O1s), and carbon (C1s) (Figure 3a). The O1s peak was attributable to adsorbed hydroxides and oxides; both specimens showed peaks at 531.9 eV. The Ti2p3/2 oxide peak at 458.5 eV was typical of Ti. The C1s peak was most often used to measure oxide levels, but the peak was weak for bare Ti. The C1s of GO featured several binding energy configurations, at 284.8, 285.1, 286.3, and 288 eV for sp2, sp3, and the C-O (epoxy/hydroxyl), and O-C5O (carboxyl) groups, respectively. The sp2 carbon (284.8 eV) was the major feature of the C1s profile, indicating the presence of GO, which was generally identified by the three characteristic Raman G, D, and 2D bands [33]. The Raman spectrum of GO showed the D band (sp3) at 1350 cm^{-1} and 1344 cm^{-1} and the G band (sp2) at 1604 cm^{-1} and 1601 cm^{-1}; bare Ti lacked these bands (Figure 3b) [34]. Figure 3c shows the XRD patterns. The typical diffraction peaks of Ti (those of the JCPDS card no. 44-1294) were observed. GO-coated Ti exhibited a broad peak at 26°, indicating between-graphene π–π stacking [35]. Hexagonal crystals of graphene or graphite were associated with characteristic peaks in the (002) and (111) planes [36]. Thus, GO clearly coated the Ti, and EPD rendered the coating uniform and thickness controllable.

Figure 4a,b shows the contact angle hydrophilicities and indentation hardness values, respectively. GO coating dramatically improved Ti hydrophilicity and hardness, reflecting the outstanding mechanical properties (Young's modulus ~1 Ta) of GO [37–39]. When metal-based implants are transplanted, strong friction and shear stresses can damage their surfaces [40]. Many coatings have been used to strengthen the surfaces [41]. Here, we simply coated GO using EPD.

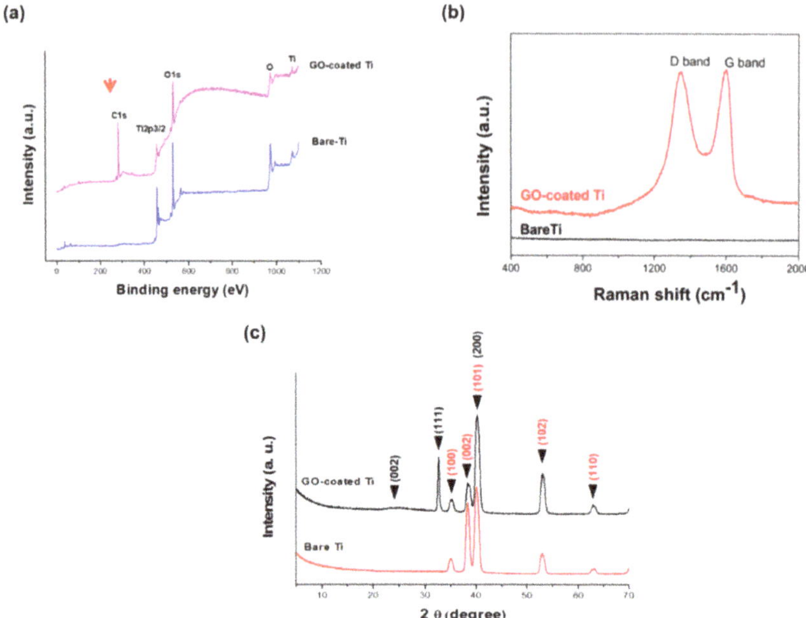

Figure 3. EPD characterization of GO-coated Ti and bare Ti. XPS spectra (**a**). Raman spectra (**b**). XRD patterns (**c**).

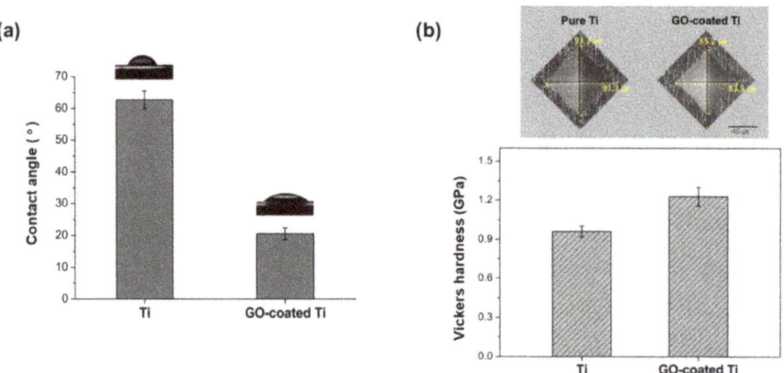

Figure 4. Contact angles (**a**) and Vickers hardness values (**b**) of bare and GO-coated Ti ($p < 0.05$).

3.3. In Vitro Cellular Responses

GO exhibited good biocompatibility and osteo-conductivity; we used CLSM to evaluate the effects of coating on stem cells, and the extents of ALP activity and mineralization compared to those of bare Ti [42–44]. Figure 5a shows CLSM images of cells cultured for 3 days. The cells were well attached, spread by 4 h, and grew over the 3 days. Neither cell attachment nor proliferation differed between the samples. To evaluate the initial (and later) osteogenic differentiation of stem cells cultured on a GO-EPD layer, cells were cultured for 7 and 14 days in a non-osteogenic culture medium. Cellular ALP activity was significantly enhanced by the GO coating. After 14 days of culture, the cellular calcium levels were measured (Figure 5c). Cells cultured on GO-coated Ti exhibited slightly more calcium deposition than those cultured on bare Ti. Not only was GO-coated Ti non-toxic but also GO facilitated early osteogenic differentiation [42–44].

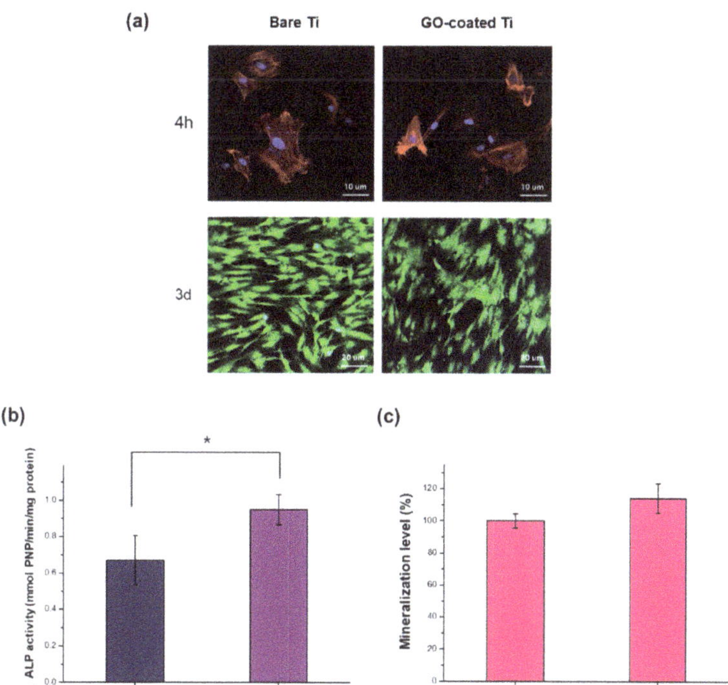

Figure 5. In vitro cell test results. Cell attachment revealed by CLSM (**a**). Alkaline phosphatase (ALP) activity of mMSCs after 7 days of culture (**b**). Alizarin Red S (ARS) staining after 21 days of culture (* $p < 0.05$) (**c**).

3.4. BM-Loading GO

GO served as both the BMP-2 loading agent and the coating. 2D flaked GO readily binds BMs; the carbon honeycomb induces BM adsorption driven by the Van der Waals force [45]. After attachment of various levels of BMP-2 to GO, BMP-2 adhesion to GO was assessed by the AFM, the Zetasizer, ELISA, and XPS. An AFM is usually employed to measure graphene thickness. As shown in Figure 6a, a blue line across the single graphene is specified and the roughness of the specimen is measured along the line. Figure 6a shows the representative AFM images of pristine GO and BM-combined GO. The AFM analysis was performed to observe the thickness change of GO combined with BM (BMP-2). The Rz value of GO and GO-GM were 4.10 ± 0.26 nm and 5.74 ± 0.61 nm, respectively. This was an increase in thickness induced by BMP-2, indicating that GO and BMP-2 were well combined.

Figure 6b shows the zeta potentials of GO with different BMP-2 concentrations; all GO coatings were deposited on the positively charged electrode. The zeta potential confirmed that EPD was in play, and that the GO and BM combination was efficient. The negative GO potential facilitated BMP-2 attachment. At a high concentration of BMP-2, the GO-BM zeta potential became more electropositive. Thus, BMP-2 adsorption to GO increased with increasing BMP-2 concentration, but the GO-BM potential remained negative; there was no need to switch the EPD anode and cathode. Figure 6c shows that at concentrations of 25 and 50 μg/mL, approximately 85% of BMP-2 became attached; the absolute concentrations of unattached BMP-2 were approximately 3.7 and 7.5 μg/mL, respectively. To minimize BMP-2 wastage, we used a BMP-2 concentration of 25 μg/mL in subsequent experiments. XPS revealed the components of the GO-BM coating (Figure 6d). The C peak and O peak of GO-coated Ti and GO-BM-coated Ti appeared at 285 and 532 eV of binding energy, respectively. Compared to GO-coated Ti, GO-BM-coated Ti (Figure 6d) exhibited a higher

N1s peak at 399 eV, attributable to the -C5N or -CN bonds caused by the N atom in the amino acid of BMP-2 [46,47].

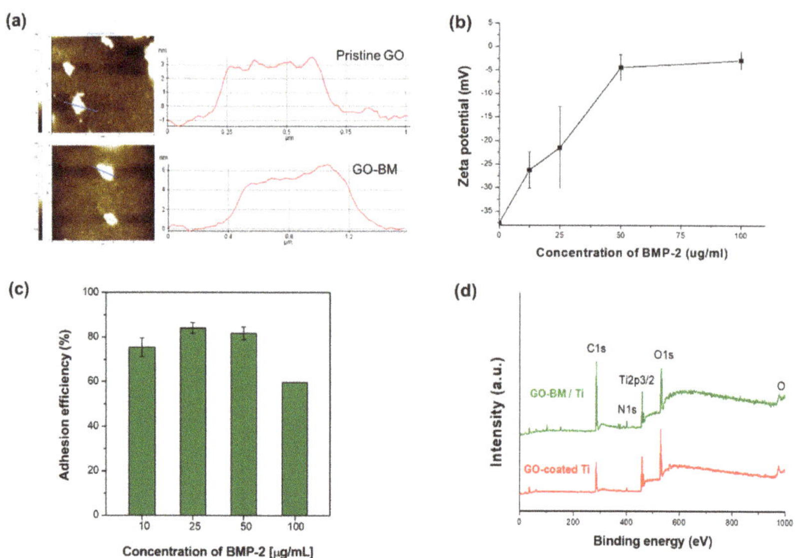

Figure 6. AFM observations of pristine GO and BM-combined GO (GO-BM) (**a**). The zeta potential of BM-combined GO as a function of the BMP-2 concentration (**b**). Characterization of GO-BM combinations by ELISA (**c**). XPS spectra of Ti with GO-BM coatings (GO-BM/Ti) and GO-coated Ti (**d**). The BM indicates BMP-2.

The GO-BM (BMP-2) coating layer was evaluated in more detail using XPS (Table 1). Ti, C, O, and N were detected in bare Ti, and the GO and GO-BM coatings; however, the N atomic ratio was highest in the latter coating. While the Ti and O proportions fell significantly in the GO-BM coating, the C proportion was higher than those of other surfaces. Most biomolecules have amine and carboxyl groups; the N and C levels were thus highest in the GO-BM coating, indicating that EPD successfully formed such a coating [48].

Table 1. XPS component analyses of coating layers.

Amount (At. %)	Bare Ti	GO	GO-BM
Ti	13.45	9.88	5.1
C	44.57	51.3	64.5
O	39.79	35.69	26.59
N	2.17	3.13	3.82

EPD forms uniform coatings. We compared GO-GFP, GO-BM, GO-coated Ti, and bare Ti. Figure 7 shows BM adhesion both photographically and as revealed by CLSM (Figure 7a–c). Compared to bare Ti, GO-coated layers had weak green fluorescence. To observe the surface of GFP-containing GO, we used adhesive tape to separate the GO-BM coating from bare Ti. The CLSM boundary data of Figure 7c show that more green fluorescence emanated from the BM-coated region. The weak green fluorescence of bare Ti may indicate that the GO-BM (GFP) suspension penetrated the adhesive tape during EPD. Unlike the surface of GO-coated Ti, a surface coated with GO-BM exhibited strong fluorescence. SEM (Figure 7d) revealed aggregates (red arrows) on the GO-BM coating. The combination of two substances during GO-BM formation was associated with aggregation or sinkage. Thus, EPD featured continuous stirring that minimized sinkage but did not completely prevent aggregation. Therefore, the green fluorescent aggregates were thought to be GO-BM complexes.

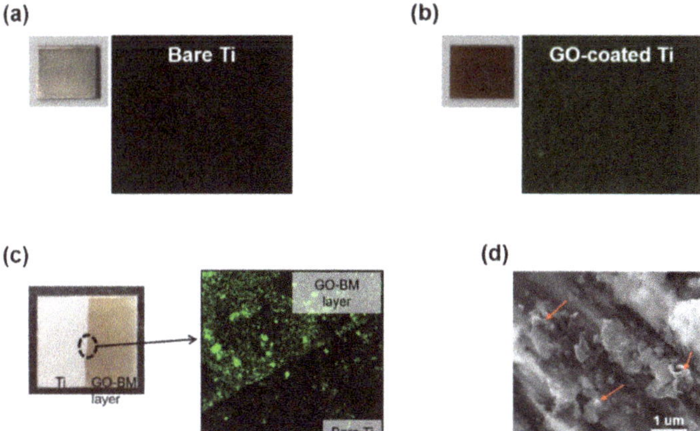

Figure 7. Optical and CLSM images of BM attachment: bare Ti (**a**), GO-coated Ti (**b**), and GO-BM/Ti (**c**). CLSM evaluation of GO-BM/Ti was performed on a region with a GO-BM layer and a region of bare Ti. SEM image of Ti with a GO-BM coating (**d**). The BM indicates GFP.

3.5. BM Release from GO-Coated Ti

BM release was assessed via CLSM and ELISA. Figure 8a shows fluorescence images of BM that remained in the GO-BM coating after soaking for various times in PBS. Over 20 days, the fluorescence intensity fell continuously, indicating sustained BM release from the GO-BM coating layer. Figure 8b shows the cumulative amounts of BM released over time. Ti exposed to BM after GO coating and Ti coated with the GO and BM combination are indicated by Post-BM and GO-BM respectively. BM was slowly and steadily released over an extended period. During up to 10 days (240 h) of analysis, both samples released similar amounts of BM. However, after 20 days, the total amounts of BM released from GO-BM/Ti and Post-BM/Ti were approximately 79.9 and 24.5 μg, respectively. CLSM revealed no significant reduction in fluorescence intensity during release up to 10 days; however, on day 14, major decreases in fluorescence intensity were evident, in line with the release profiles. After 10 days, BM was no longer released from the Post-BM/Ti sample; however, GO-BM/Ti then exhibited continued rapid BM release, unlike the previous steady release profile.

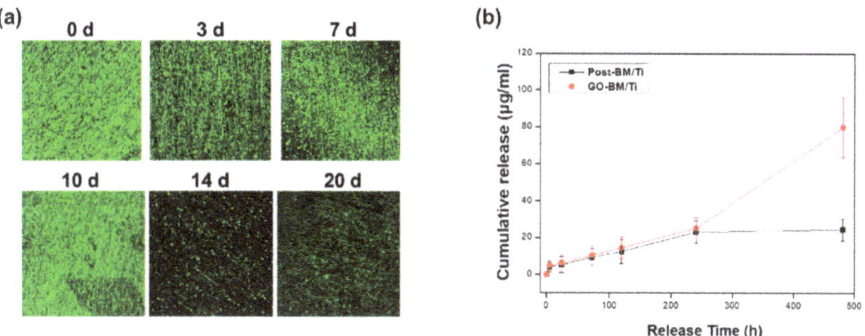

Figure 8. Release behaviors of BM (GFP). CLSM fluorescence images of BM remaining in GO-BM/Ti (**a**). BM release profiles over time (**b**).

We found that BM pervaded the coating, and we demonstrated how to modify metal implants to ensure stable long-term BM release. The coating thickness controlled the amount and rate of BM release. Room temperature EPD coated undamaged BMs; it was easy to adjust the coating thickness. GO readily adsorbed BMs. High-quality coatings of varying (controllable) thickness formed rapidly. BM loadings were high, because the BMs were not (only) surface-attached.

4. Conclusions

We used EPD to modify metal surfaces to stably and slowly deliver BMs. GO coating increased surface hydrophilicity and hardness and ALP activity (a feature of osteogenic differentiation). Hydrophilic GO combined with BMs to form GO-BM complexes that uniformly coated a metal. Coatings with internal (not only surface-attached) BMs were optimal, and will find many applications in medicine.

Author Contributions: Conceptualization, E.-J.L.; methodology, J.-S.O. and J.-H.J.; investigation, E.-J.L., J.-S.O. and J.-H.J.; writing—original draft preparation, E.-J.L., J.-S.O. and J.-H.J.; writing—review and editing, E.-J.L., J.-S.O. and J.-H.J.; supervision, E.-J.L.; project administration, E.-J.L.; funding acquisition, E.-J.L. All authors have read and agreed to the published version of the manuscript.

Funding: This work was supported by the National Research Foundation of Korea (NRF), which is funded by the Ministry of Science and Information and Communications Technologies (MSIT; NRF-2018M3C1B7021994 and NRF-2020R1A2C1012454).

Data Availability Statement: Not applicable.

Acknowledgments: This work was supported by the National Research Foundation of Korea (NRF), which is funded by the Ministry of Science and Information and Communications Technologies (MSIT; NRF-2018M3C1B7021994 and NRF-2020R1A2C1012454).

Conflicts of Interest: The author(s) declared no potential conflicts of interest with respect to the research, authorship, and/or publication of this article.

References

1. Alvarez, K.; Nakajima, H. Metallic scaffolds for bone regeneration. *Materials* **2009**, *2*, 790–832. [CrossRef]
2. Pizzoferrato, A.; Cenni, E.; Ciapetti, G.; Granchi, D.; Savarino, L.; Stea, S. Inflammatory response to metals and ceramics. In *Integrated Biomaterials Science*; Barbucci, R., Ed.; Springer US: Boston, MA, USA, 2002; pp. 735–791.
3. Shahali, H.; Jaggessar, A.; Yarlagadda, P.K.D.V. Recent advances in manufacturing and surface modification of titanium orthopaedic applications. *Procedia Eng.* **2017**, *174*, 1067–1076. [CrossRef]
4. Dong, H.; Liu, H.; Zhou, N.; Li, Q.; Yang, G.; Chen, L.; Mou, Y. Surface modified techniques and emerging functional coating of dental implants. *Coatings* **2020**, *10*, 1012. [CrossRef]
5. Kim, Y.-W. Surface modification of ti dental implants by grit-blasting and micro-arc oxidation. *Mater. Manuf. Process.* **2010**, *25*, 307–310. [CrossRef]
6. López, V.; Sundaram, R.S.; Gómez-Navarro, C.; Olea, D.; Burghard, M.; Gómez-Herrero, J.; Zamora, F.; Kern, K. Chemical vapor deposition repair of graphene oxide: A route to highly-conductive graphene monolayers. *Adv. Mater.* **2009**, *21*, 4683–4686. [CrossRef]
7. Vats, N.; Rauschenbach, S.; Sigle, W.; Sen, S.; Abb, S.; Portz, A.; Dürr, M.; Burghard, M.; van Aken, P.A.; Kern, K. Electron microscopy of polyoxometalate ions on graphene by electrospray ion beam deposition. *Nanoscale* **2018**, *10*, 4952–4961. [CrossRef] [PubMed]
8. Terasawa, T.-O.; Saiki, K. Growth of graphene on cu by plasma enhanced chemical vapor deposition. *Carbon* **2012**, *50*, 869–874. [CrossRef]
9. Wu, C.; Cheng, Q.; Sun, S.; Han, B. Templated patterning of graphene oxide using self-assembled monolayers. *Carbon* **2012**, *50*, 1083–1089. [CrossRef]
10. Beutner, R.; Michael, J.; Schwenzer, B.; Scharnweber, D. Biological nano-functionalization of titanium-based biomaterial surfaces: A flexible toolbox. *J. R. Soc. Interface* **2010**, *7*, S93–S105. [CrossRef] [PubMed]
11. Haimov, H.; Yosupov, N.; Pinchasov, G.; Juodzbalys, G. Bone morphogenetic protein coating on titanium implant surface: A systematic review. *J. Oral. Maxillofac. Res.* **2017**, *8*, e1. [CrossRef]
12. Ajori, S.; Ameri, A.; Ansari, R. Adsorption analysis and mechanical characteristics of carbon nanotubes under physisorption of biological molecules in an aqueous environment using molecular dynamics simulations. *Mol. Simul.* **2020**, *46*, 388–397. [CrossRef]

13. Ouyang, L.; Deng, Y.; Yang, L.; Shi, X.; Dong, T.; Tai, Y.; Yang, W.; Chen, Z.-G. Graphene-oxide-decorated microporous polyetheretherketone with superior antibacterial capability and in vitro osteogenesis for orthopedic implant. *Macromol. Biosci.* **2018**, *18*, 1800036. [CrossRef]
14. Zhao, C.; Lu, X.; Zanden, C.; Liu, J. The promising application of graphene oxide as coating materials in orthopedic implants: Preparation, characterization and cell behavior. *Biomed. Mater.* **2015**, *10*, 015019. [CrossRef] [PubMed]
15. Hirata, E.; Ménard-Moyon, C.; Venturelli, E.; Takita, H.; Watari, F.; Bianco, A.; Yokoyama, A. Carbon nanotubes functionalized with fibroblast growth factor accelerate proliferation of bone marrow-derived stromal cells and bone formation. *Nanotechnology* **2013**, *24*, 435101. [CrossRef] [PubMed]
16. La, W.-G.; Park, S.; Yoon, H.-H.; Jeong, G.-J.; Lee, T.-J.; Bhang, S.H.; Han, J.Y.; Char, K.; Kim, B.-S. Delivery of a therapeutic protein for bone regeneration from a substrate coated with graphene oxide. *Small* **2013**, *9*, 4051–4060. [CrossRef] [PubMed]
17. Oh, J.-S.; Lee, E.-J. Enhanced effect of polyethyleneimine-modified graphene oxide and simvastatin on osteogenic differentiation of murine bone marrow-derived mesenchymal stem cells. *Biomedicines* **2021**, *9*, 501. [CrossRef]
18. Wu, J.; Zheng, A.; Liu, Y.; Jiao, D.; Zeng, D.; Wang, X.; Cao, L.; Jiang, X. Enhanced bone regeneration of the silk fibroin electrospun scaffolds through the modification of the graphene oxide functionalized by bmp-2 peptide. *Int. J. Nanomed.* **2019**, *14*, 733–751. [CrossRef]
19. Saladino, M.L.; Markowska, M.; Carmone, C.; Cancemi, P.; Alduina, R.; Presentato, A.; Scaffaro, R.; Biały, D.; Hasiak, M.; Hreniak, D.; et al. Graphene oxide carboxymethylcellulose nanocomposite for dressing materials. *Materials* **2020**, *13*, 1980. [CrossRef]
20. Li, H.; Gao, C.; Tang, L.; Wang, C.; Chen, Q.; Zheng, Q.; Yang, S.; Sheng, S.; Zan, X. Lysozyme (lys), tannic acid (ta), and graphene oxide (go) thin coating for antibacterial and enhanced osteogenesis. *ACS Appl. Bio Mater.* **2020**, *3*, 673–684. [CrossRef]
21. Han, L.; Sun, H.; Tang, P.; Li, P.; Xie, C.; Wang, M.; Wang, K.; Weng, J.; Tan, H.; Ren, F.; et al. Mussel-inspired graphene oxide nanosheet-enwrapped ti scaffolds with drug-encapsulated gelatin microspheres for bone regeneration. *Biomater. Sci.* **2018**, *6*, 538–549. [CrossRef]
22. Oliver, J.-A.N.; Su, Y.; Lu, X.; Kuo, P.-H.; Du, J.; Zhu, D. Bioactive glass coatings on metallic implants for biomedical applications. *Bioact. Mater.* **2019**, *4*, 261–270. [CrossRef]
23. Santoro, M.; Tatara, A.M.; Mikos, A.G. Gelatin carriers for drug and cell delivery in tissue engineering. *J. Control. Release* **2014**, *190*, 210–218. [CrossRef] [PubMed]
24. Fox, K.E.; Tran, N.L.; Nguyen, T.A.; Nguyen, T.T.; Tran, P.A. 8—surface modification of medical devices at nanoscale—recent development and translational perspectives. In *Biomaterials in Translational Medicine*; Yang, L., Bhaduri, S.B., Webster, T.J., Eds.; Academic Press: Cambridge, MA, USA, 2019; pp. 163–189.
25. Besra, L.; Liu, M. A review on fundamentals and applications of electrophoretic deposition (epd). *Prog. Mater. Sci.* **2007**, *52*, 1–61. [CrossRef]
26. Ma, Y.; Han, J.; Wang, M.; Chen, X.; Jia, S. Electrophoretic deposition of graphene-based materials: A review of materials and their applications. *J. Mater.* **2018**, *4*, 108–120. [CrossRef]
27. Mallick, M.; Arunachalam, N. Electrophoretic deposited graphene based functional coatings for biocompatibility improvement of nitinol. *Thin Solid Films* **2019**, *692*, 137616. [CrossRef]
28. Eshghinejad, P.; Farnoush, H.; Bahrami, M.S.; Bakhsheshi-Rad, H.R.; Karamian, E.; Chen, X.B. Electrophoretic deposition of bioglass/graphene oxide composite on ti-alloy implants for improved antibacterial and cytocompatible properties. *Mater. Technol.* **2020**, *35*, 69–74. [CrossRef]
29. Mehrali, M.; Akhiani, A.R.; Talebian, S.; Mehrali, M.; Latibari, S.T.; Dolatshahi-Pirouz, A.; Metselaar, H.S.C. Electrophoretic deposition of calcium silicate–reduced graphene oxide composites on titanium substrate. *J. Eur. Ceram. Soc.* **2016**, *36*, 319–332. [CrossRef]
30. Shi, Y.Y.; Li, M.; Liu, Q.; Jia, Z.J.; Xu, X.C.; Cheng, Y.; Zheng, Y.F. Electrophoretic deposition of graphene oxide reinforced chitosan–hydroxyapatite nanocomposite coatings on ti substrate. *J. Mater. Sci. Mater. Med.* **2016**, *27*, 48. [CrossRef] [PubMed]
31. Sharma, H.; Mondal, S. Functionalized graphene oxide for chemotherapeutic drug delivery and cancer treatment: A promising material in nanomedicine. *Int. J. Mol. Sci.* **2020**, *21*, 6280. [CrossRef]
32. Rykowska, I.; Nowak, I.; Nowak, R. Drug-eluting stents and balloons-materials, structure designs, and coating techniques: A review. *Molecules* **2020**, *25*, 4624. [CrossRef]
33. Kovtun, A.; Jones, D.; Dell'Elce, S.; Treossi, E.; Liscio, A.; Palermo, V. Accurate chemical analysis of oxygenated graphene-based materials using x-ray photoelectron spectroscopy. *Carbon* **2019**, *143*, 268–275. [CrossRef]
34. Qin, H.; Gong, T.; Cho, Y.; Lee, C.; Kim, T. A conductive copolymer of graphene oxide/poly(1-(3-aminopropyl)pyrrole) and the adsorption of metal ions. *Polym. Chem.* **2014**, *5*, 4466–4473. [CrossRef]
35. Karthik, P.; Vinoth, R.; Zhang, P.; Choi, W.; Balaraman, E.; Neppolian, B. Π–π interaction between metal–organic framework and reduced graphene oxide for visible-light photocatalytic h2 production. *ACS Appl. Energy Mater.* **2018**, *1*, 1913–1923. [CrossRef]
36. Xu, X.; Li, W.; Wang, Y.; Dong, G.; Jing, S.; Wang, Q.; Feng, Y.; Fan, X.; Ding, H. Study of the preparation of cu-tic composites by reaction of soluble ti and ball-milled carbon coating tic. *Results Phys.* **2018**, *9*, 486–492. [CrossRef]
37. Hwang, M.-J.; Kim, M.-G.; Kim, S.; Kim, Y.C.; Seo, H.W.; Cho, J.K.; Park, I.-K.; Suhr, J.; Moon, H.; Koo, J.C.; et al. Cathodic electrophoretic deposition (epd) of phenylenediamine-modified graphene oxide (go) for anti-corrosion protection of metal surfaces. *Carbon* **2019**, *142*, 68–77. [CrossRef]

38. Zielinski, A.; Bartmanski, M. Electrodeposited biocoatings, their properties and fabrication technologies: A review. *Coatings* **2020**, *10*, 782. [CrossRef]
39. Nemee, P.; Jaitanong, N.; Narksitipan, S. Surface modification of low carbon steel by using electrophoretic deposition technique with graphene oxide powder. *Solid State Phenom.* **2020**, *302*, 1–7. [CrossRef]
40. Shayesteh Moghaddam, N.; Taheri Andani, M.; Amerinatanzi, A.; Haberland, C.; Huff, S.; Miller, M.; Elahinia, M.; Dean, D. Metals for bone implants: Safety, design, and efficacy. *Biomanuf. Rev.* **2016**, *1*, 1. [CrossRef]
41. Dehghanghadikolaei, A.; Fotovvati, B. Coating techniques for functional enhancement of metal implants for bone replacement: A review. *Materials* **2019**, *12*, 1795. [CrossRef]
42. Su, J.; Du, Z.; Xiao, L.; Wei, F.; Yang, Y.; Li, M.; Qiu, Y.; Liu, J.; Chen, J.; Xiao, Y. Graphene oxide coated titanium surfaces with osteoimmunomodulatory role to enhance osteogenesis. *Mater. Sci. Eng. C* **2020**, *113*, 110983. [CrossRef]
43. Qiu, J.; Geng, H.; Wang, D.; Qian, S.; Zhu, H.; Qiao, Y.; Qian, W.; Liu, X. Layer-number dependent antibacterial and osteogenic behaviors of graphene oxide electrophoretic deposited on titanium. *ACS Appl. Mater. Interfaces* **2017**, *9*, 12253–12263. [CrossRef] [PubMed]
44. Tao, B.; Chen, M.; Lin, C.; Lu, L.; Yuan, Z.; Liu, J.; Liao, Q.; Xia, Z.; Peng, Z.; Cai, K. Zn-incorporation with graphene oxide on ti substrates surface to improve osteogenic activity and inhibit bacterial adhesion. *J. Biomed. Mater. Res. A* **2019**, *107*, 2310–2326. [CrossRef]
45. Allen, M.J.; Tung, V.C.; Kaner, R.B. Honeycomb carbon: A review of graphene. *Chem. Rev.* **2010**, *110*, 132–145. [CrossRef]
46. Song, M.-J.; Amirian, J.; Linh, N.T.B.; Lee, B.-T. Bone morphogenetic protein-2 immobilization on porous pcl-bcp-col composite scaffolds for bone tissue engineering. *J. Appl. Polym. Sci.* **2017**, *134*, 45186. [CrossRef]
47. Pugalenthi, R.; Arunachalam, P.; Amalraj, S.; Rathinavel, J.; Subramaniyan, D.S.; Bhagavathsingh, J.; Samuel, V. Covalent grafting of n-containing compound with graphene oxide: Efficient electrode material for supercapacitor. *Chem. Sci. Eng. Res.* **2019**, *1*. [CrossRef]
48. Liang, X.; Zhong, J.; Shi, Y.; Guo, J.; Huang, G.; Hong, C.; Zhao, Y. Hydrothermal synthesis of highly nitrogen-doped few-layer graphene via solid–gas reaction. *Mater. Res. Bull.* **2015**, *61*, 252–258. [CrossRef]

MDPI

Article

Characterization of Titanium Surface Modification Strategies for Osseointegration Enhancement

Jinyoung Kim [1,†], Hyun Lee [2,3,†], Tae-Sik Jang [4], DongEung Kim [5], Chang-Bun Yoon [6], Ginam Han [2,3], Hyoun-Ee Kim [1] and Hyun-Do Jung [2,3,*]

1 Department of Materials Science and Engineering, Seoul National University, Seoul 08826, Korea; tmr002@snu.ac.kr (J.K.); kimhe@snu.ac.kr (H.-E.K.)
2 Department of BioMedical-Chemical Engineering (BMCE), The Catholic University of Korea, Bucheon 14662, Korea; leeh0520@catholic.ac.kr (H.L.); gnhan@catholic.ac.kr (G.H.)
3 Department of Biotechnology, The Catholic University of Korea, Bucheon 14662, Korea
4 Department of Materials Science and Engineering, Chosun University, Gwangju 61452, Korea; tsjang@chosun.ac.kr
5 Research Institute of Advanced Manufacturing Technology, Korea Institute of Industrial Technology, Incheon 21999, Korea; canon@kitech.re.kr
6 Department of Advanced Materials Engineering, Korea Polytechnic University, Siheung 15073, Korea; cbyoon@kpu.ac.kr
* Correspondence: hdjung@catholic.ac.kr; Tel.: +82-2-2164-4357
† Jinyoung Kim and Hyun Lee have contributed equally to this work.

Abstract: As biocompatible metallic materials, titanium and its alloys have been widely used in the orthopedic field due to their superior strength, low density, and ease of processing. However, further improvement in biological response is still required for rapid osseointegration. Here, various Ti surface-treatment technologies were applied: hydroxyapatite blasting, sand blasting and acid etching, anodic oxidation, and micro-arc oxidation. The surface characteristics of specimens subjected to these techniques were analyzed in terms of structure, elemental composition, and wettability. The adhesion strength of the coating layer was also assessed for the coated specimens. Biocompatibility was compared via tests of in vitro attachment and proliferation of pre-osteoblast cells.

Keywords: titanium; surface treatment; HA blasting; sandblasted and acid-etched (SLA); anodic oxidation (AO); micro-arc oxidation (MAO)

Citation: Kim, J.; Lee, H.; Jang, T.-S.; Kim, D.; Yoon, C.-B.; Han, G.; Kim, H.-E.; Jung, H.-D. Characterization of Titanium Surface Modification Strategies for Osseointegration Enhancement. *Metals* **2021**, *11*, 618. https://doi.org/10.3390/met11040618

Academic Editor: Francesca Borgioli

Received: 6 March 2021
Accepted: 8 April 2021
Published: 11 April 2021

1. Introduction

Various materials have been used as biomaterials to replace damaged organs or tissues inside the human body. Different materials have been chosen for application to different damaged body parts based on the intrinsic properties of each material. Such materials are classified into three main groups: metals, ceramics, and polymers. Metals and ceramics have been applied mostly to hard tissue engineering, because its inherent mechanical properties are sufficient to endure applied loads [1,2]. By contrast, polymers have been broadly utilized in soft and hard tissue engineering, since their properties are determined by not only molecular weight and fabrication method but also constituent elements and chain structure [3,4]. Among these, metallic materials have been more widely used as hard tissue substitutes than ceramics or polymers because of their excellent mechanical properties, chemical stability in physiological conditions, and ductility, which can prevent sudden fracture [5,6]. Given biocompatibility considerations, the most prevalently facilitated metallic materials have been stainless steel (SUS) [7], Co–Cr alloys [8], magnesium (Mg) [9], and titanium and its alloys [10–13].

Titanium and its alloys are the most extensively utilized metals because of their excellent biocompatibility, high strength-to-weight ratios, and ease of processing, in addition to the characteristics of metals mentioned earlier [5,14,15]. Unlike other biologically compatible metals, Ti and Ti alloys exhibit exceptional corrosion resistance due to the formation of a passive oxide film layer [16]. This TiO_2 layer, bonded tightly to Ti, greatly increases resistance to all types of corrosion while simultaneously improving cellular compatibility [5,14]. Furthermore, Ti alloys including Ti–6Al–4V [17,18], Ti–Nb–Zr [19,20], and Ti–Nb–Zr–Ta alloys [21,22] have been developed for increased strength, enhanced biocompatibility, and reduced elastic modulus, respectively, using theoretical calculations to consider the effect of each element (DV-Xα) [23,24]. The effectiveness of each of these Ti alloys has already been studied and demonstrated [25–28].

Even though Ti and Ti alloys exhibit suitable properties for utilization as biomaterials, there remains a potential improvement in biocompatibility for long-term stability. Fast-paced research aimed at solving these problems through surface treatment of Ti and Ti alloys is already under way [29–32]. One commonly implemented method is to increase osseointegration by surface coating with various materials such as calcium phosphate for improved biocompatibility. In another method, micro- or nano-level roughness is added to the surface structure, and an oxide layer is intentionally created to maximize the effect. Increasing the surface roughness improves cell characteristics [33,34] and the stability of the implant [35], and it can be used for drug loading or as an antibacterial coating layer [36–38]. The formation of a calcium phosphate-based coating layer on titanium for various purposes has been widely studied [5,39]. It is known that the sand blasting and acid etching (SLA) method, which is the most often used strategy to alter the surface structure of titanium, not only increases biometric characteristics by increasing the roughness [40] but also is suitable for implants requiring mechanical strength [41]. In some studies, however, SLA specimens exhibited poorer cellular properties than those of specimens with a smooth surface [42]. Furthermore, granules and etching agents that were not completely removed after SLA treatment could affect the osteointegration of the implant [43]. Therefore, in addition to SLA, a method such as target ion-induced plasma sputtering can be used to increase the biocompatibility by increasing the roughness [44], and the structure of the surface layer can thus be made more suitable for implants through anodic oxidation (AO) and micro arc oxidation (MAO) methods that use electrical treatment to intentionally create an oxide layer [45,46].

In this study, HA blasting and SLA was firstly applied to prepare surface-treated Ti substrates (noted as HA and SLA). Then AO and MAO processes were additionally conducted on the SLA-treated Ti specimens to form a supplementary titanium dioxide layer (noted as SLA/AO and SLA/MAO). The physiological and biological properties were compared through the following methods. The microstructure of the surface and the coating thickness were assessed through field-emission scanning electron microscopy (FE-SEM) with a focused ion beam (FIB). Elemental alteration was confirmed through X-ray diffraction (XRD) and energy-dispersive X-ray spectroscopy (EDS). Furthermore, the roughness of the surface-treated Ti specimens was evaluated by laser scanning microscopy (LSM), and the hydrophilicity of specimens was also measured. Coating layers were formed using a sequence of hydroxyapatite (HA) blasting, AO, and MAO, and the stability of each layer was estimated. In vitro tests of attachment and proliferation were conducted to compare the effectiveness of various surface-treatment techniques.

2. Materials and Methods

2.1. Preparation of Surface-Treated Ti Substrates

Commercially available Ti plates (Commercially pure Ti G4) were prepared with dimensions of 10 mm × 10 mm × 2 mm. Ti plates were sequentially polished using P400, P1000, and P2000 SiC papers. After polishing, Ti plates were washed in ethanol under sonication to remove debris and then dried at a temperature below 70 °C prior to surface treatment. In the preparation of HA-blasted Ti plates (hereafter HA), Ti plates were impacted by jetted HA particles and then rinsed with nitric acid to eliminate remaining HA sandblasting particles. In the case of SLA treatment, an identical HA-blasting procedure was conducted prior to the following post-treatment. Nitric acid-treated Ti plates were immersed in a 5 M NaOH solution for 24 h at 60 °C and then soaked in distilled water for 24 h at 80 °C. Specimens were then dried in an oven overnight at 40 °C, followed by heat treatment at 600 °C for 1 h of dwelling time. AO and MAO procedures were applied to form a titanium oxide (TiO_2) layer on the SLA-Ti surface. The MAO process used an electrolytic solution containing 0.25 wt% ammonium fluoride (Sigma-Aldrich, Saint Louis, MO, USA), whereas the AO process used a solution of 98 vol% ethylene glycol (Sigma-Aldrich) and 2 vol% distilled water. SLA-treated Ti plates were dipped in the prepared electrolytic solution, and a DC field (HD-9001D, FinePower, Seoul, Korea) of 60 V was applied for 30 min. Fabricated SLA/AO-Ti plates were ultrasonically cleaned sequentially in acetone, ethanol, and distilled water. MAO treatment was performed in an electrolytic aqueous solution of 0.15 M calcium acetate monohydrate (Sigma-Aldrich) and 0.02 M glycerol phosphate calcium salt (Sigma-Aldrich). Electric power from a pulsed DC field (Pulse Power Supply; Model-P6241, Auto Electric Co., Seoul, Korea) was supplied to the specimens and the stainless steel (SUS) counter electrode for 2 min under the following conditions: voltage, 350 V; frequency, 660 Hz; and 60% duty. A cooling bath at 10 °C was utilized to prevent temperature rise during the process. After treatment, SLA/MAO-Ti plates were again rinsed with distilled water and ethanol.

2.2. *Characterization of Ti Substrates after Surface Treatments*

The surface morphologies of prepared surface-treated Ti plates were analyzed using FE-SEM (JSM-6330F, JEOL, Tokyo, Japan). For each of HAs, SLA/AO, and SLA/MAO, the structure and thickness of the coating layer were observed using a FIB (AURIGA, Carl Zeiss, Oberkochen, Germany) to section the coating layer. An elemental analysis was conducted by EDS equipped with FE-SEM, and the crystalline phases of each specimen were analyzed using XRD (D8-advance, BRUKER Co., Billerica, MA, USA) with 20–80° of scanning range and a 1°/min scanning rate. Additionally, the hydrophilicity of surface-treated Ti plates was assessed by measuring the aqueous wetting angle using a contact angle analyzer (Phoenix 300; Surface Electro Optics Co., Suwon, Korea). In total, 10 μL of distilled water droplets were dropped onto each surface from a syringe, and the morphology of the water droplet was monitored at particular time intervals: 0 min, 2 min, 6 min, and 10 min. The wetting angle was estimated by using an image analysis program (Image XP). The surface roughness of prepared surface-treated Ti plates was analyzed by LSM (OLS 3100, Olympus, Tokyo, Japan). The surface root mean square height (Sq) was determined for a surface area of 1280×1280 μm^2 using a 5× objective lens at a wavelength of 408 nm. The adhesive strength between the coating layer and the Ti substrate was measured using the pull-out test, and three specimens for each condition were examined. Aluminum studs with pre-epoxy coating were attached to each surface and heated at 150 °C to cure the epoxy. Studs attached to each coating were pulled until detachment by a universal testing machine (RB302 single column type; R&B, Daejeon, Korea) at a rate of 1 mm/min. The remaining coating layers were observed by FE-SEM after detachment.

2.3. *In Vitro Biocompatibility Assessment*

In vitro cell tests, including attachment and proliferation tests, were carried out to assess the biocompatibility of treated Ti substrates. Sequential sterilization by immersion in 70%EtOH, autoclaving at 121 °C, and UV irradiation were performed on all of the specimens before the tests. Pre-osteoblast cells (MC3T3-E1; ATCC, CRL-2593, Manassas, VA, USA) were cultured in the alpha-minimum essential medium (α-MEM; Welgene Co., Gyeongsan, Korea) containing 10% fetal bovine serum (FBS; Gibco, New York, NY, USA) and 1% penicillin–streptomycin (Pen–strep; Gibco) in a humidified incubator at 37 °C and a CO_2 concentration below 5%. For cell attachment assessment, prepared specimens were immersed in a medium containing 3×10^4 cells per mL. After 3 h of culturing, adhered cells were fixed by 2.5% glutaraldehyde (Sigma-Aldrich) and dehydrated using 75%, 95%, and 100% EtOH, as well as 1,1,1,3,3,3-hexamethyldisilazane (Sigma-Aldrich). The morphology of cells was observed by FE-SEM. The degree of cell proliferation was measured after the methoxyphenyl tetrazolium salt (MTS) method was carried out. A total of 3×10^4 cells/mL were cultured on the treated Ti substrates for 3 days and 5 days. Three specimens of each group were tested. For each incubation time, specimens were rinsed twice with DPBS (Dulbecco's phosphate buffered saline; Welgene Co.) to remove non-attached cells. The specimens were then immersed in FBS-free α-MEM containing a 10% cell proliferation assay kit solution (CellTiter 96 Aqueous One Solution Cell Proliferation Assay) and cultivated in an incubator to induce sufficient reaction. Afterward, the absorbance of the medium was measured using a microplate reader (EX read 400, Biochrom, Hollistone, MA, UK) at a wavelength of 490 nm.

2.4. Statistical Analysis

Statistical analysis was carried out using statistical package for the social sciences (SPSS 27, SPSS Inc., Chicago, IL, USA). One-way analysis of variance with Tukey post hoc was performed to the results of adhesion strength, and Kruskal-Wallis H test with pairwise comparison post hoc was conducted on the cell proliferation data. The p values less than 0.05 were considered statistically significant.

3. Results and Discussion

Figure 1 illustrates FE-SEM observations of the surface structures of surface-treated Ti specimens. The unique polishing pattern is clearly visible in the Ti specimen (Figure 1a). By contrast, the HA-blasted specimens and SLAs in Figure 1b,c show surfaces roughened by the blasting and acid treatment process. The highly magnified image indicates a sharply carved surface morphology derived from blasting and etching. Contrarily, unique structures were observed in the two experimental groups that underwent additional surface treatment through electro-oxidation. As demonstrated in Figure 1d, considerable differences were not found between SLA/AO specimens and HA and SLA specimens based on low-magnification images. However, as indicated in the inset image, nanoporous rods were generated on the SLA-treated Ti surface; this phenomenon is typical in AO treatment [34,44]. Meanwhile, trace of SLA treatment could not be seen in the SLA/MAO specimens (Figure 1e), and a typical crater-like porous structure was formed on the surface. This structure was generated by the numerous arcs on the Ti surface during the MAO treatment process, and the structure and its shapes are consistent with structures documented in various MAO studies [47,48]. This finding suggests that the additional treatment to SLA-treated Ti substrates could induce a more roughened structure or formation of a coating layer.

Unlike the SLA-treated specimens, which removed foreign substances from the surface through an acid treatment process, the HA samples and the samples treated with AO and MAO after SLA exhibited coating layers with different thicknesses. Coating layer thicknesses for HA-, SLA/AO-, and SLA/MAO-treated Ti surfaces were identified through FIB (Figure 2). Since surface treatments were performed in optimized conditions with effectiveness, the thicknesses of the formed coating layers differed. The average measured thickness for each group was 6 μm, 2 μm, and 1 μm for SLA/MAO, HA, and SLA/AO, respectively. According to Wu and Kuromoto, the thickness of the coating layer of AO changes according to the anodizing time and volume [49,50]; conditions applied here yielded a thickness of approximately 1 μm. MAO-treated specimens exhibited a thicker coating layer than AO-treated ones because the applied bias voltage was much higher in MAO treatment (350 V) than that in AO treatment (60 V). This difference resulted in a far more intense electric reaction on the Ti surface in the electrolyte for MAO than that for AO.

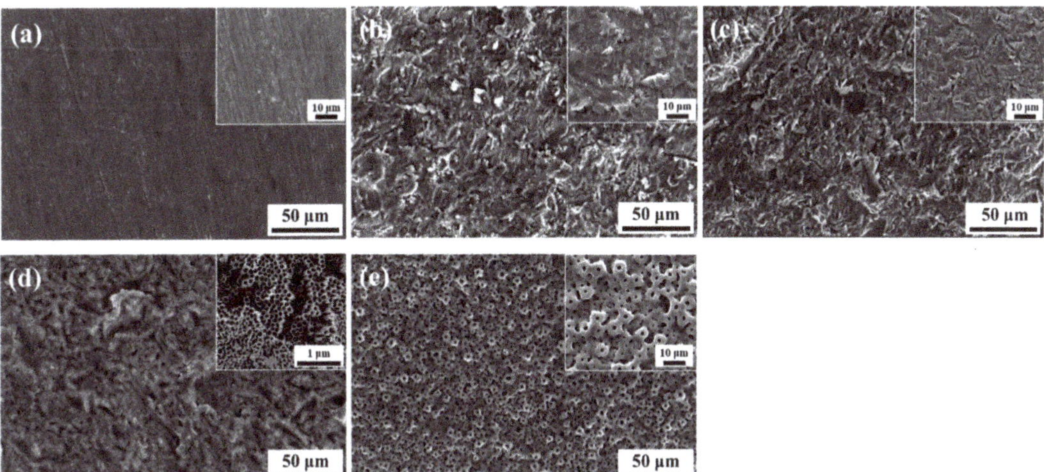

Figure 1. Surface morphology of surface-modified Ti through various treatments: (**a**) Ti, (**b**) hydroxyapatite blasting (HA), (**c**) sandblasted and acid-etched (SLA), (**d**) SLA/anodic oxidation (AO), and (**e**) SLA/micro arc oxidation (MAO).

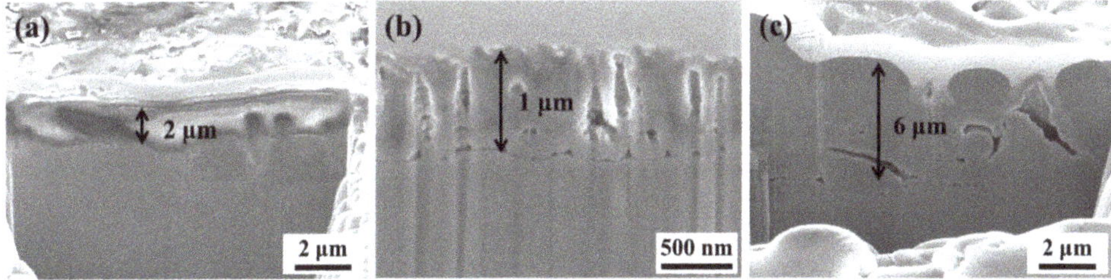

Figure 2. Focused ion beam (FIB) images of surface-modified Ti: (**a**) HA, (**b**) SLA/AO, and (**c**) SLA/MAO.

The elemental characteristics of each experimental condition were analyzed through XRD and EDS analyses. Figure 3a illustrates the crystalline phases of each Ti substrate after surface treatment. Compared to the crystalline phases of pristine Ti, SLA-treated Ti specimens exhibited identical results, since all of the undesired residues were eliminated after SLA treatment [40]. By contrast, HA-Ti specimens exhibited representative HA peaks because HA particles were embedded into the Ti substrate. Specimens treated by electrical discharge exhibited peaks related to anatase TiO_2. However, in SLA/AO, the TiO_2 peak was not clearly observed because the TiO_2 formed through anodizing was in an amorphous state [34]. Furthermore, Figure 3b exhibits embedded elements on the specimen surfaces, and only HA- and SLA/MAO-treated Ti surfaces exhibited evidence of the Ca and P elements. These Ca and P peaks emerged because oxidation by the generated electric arc caused HA particles (HA treated Ti) to be embedded, incorporating Ca and P in the TiO_2 coating layer (SLA/MAO-treated Ti) [51,52].

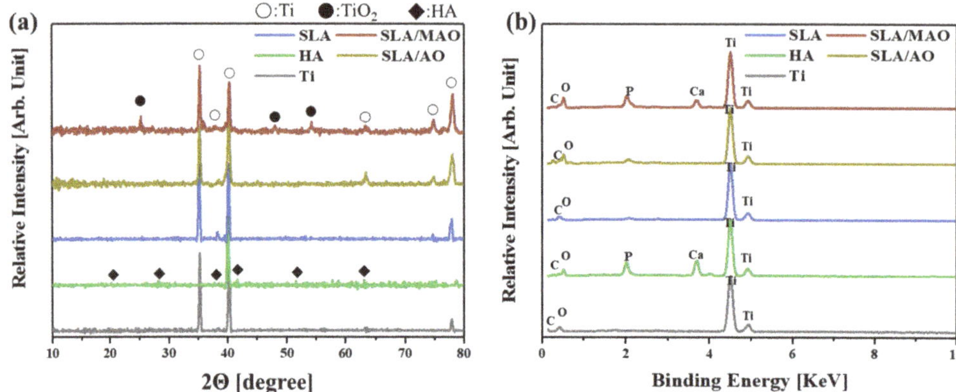

Figure 3. Component analysis of surface-modified titanium through (**a**) X-ray diffraction (XRD) and (**b**) energy-dispersive X-ray spectroscopy (EDS) analyses.

For good osseointegration between the artificial implant and the surrounding bones, a roughened surface is required, since the initial adhesion of osteoblasts is thereby improved; surrounding irregularities promote the differentiation of stem cells into bone cells [53]. Since the experimental methods for treating Ti surfaces in this study have been acknowledged as useful ways of creating rough surfaces, a quantitative roughness assessment was conducted. Figure 4 presents LSM images of treated surfaces; Table 1 summarizes the LSM-measured Sq values. As illustrated in Figure 4, the pristine Ti exhibited an almost flat state. The SLA specimen exhibited minimal roughness. All of the experimental groups displayed a significantly higher degree of roughness than pristine Ti. HA specimens (Figure 4b) were much rougher because of the carving and fixation of HA particles resulting from collisions between HA particles and the Ti surface. SLA-treated specimens showed higher S_q value than pure Ti as illustrated in Figure 4c with evenly distributed color map. The SLA/AO specimens indicated slightly higher roughness than SLA group because of the generated topmost nanoporous TiO_2 layer. The most roughened surfaces were specimens treated using SLA/MAO, as is clear from the drastic color distribution in Figure 4e. A highly drastic reaction occurring on the Ti surface induced an irregularly distributed TiO_2 coating layer with crater-like micropores.

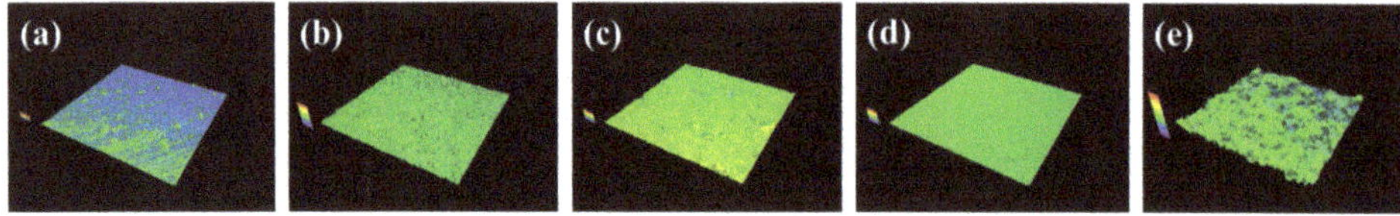

Figure 4. Roughness of surface-modified Ti: (**a**) Ti, (**b**) HA, (**c**) SLA, (**d**) SLA/AO, and (**e**) SLA/MAO.

Table 1. S_q value of surface-modified Ti.

Specimens	Ti	HA	SLA	SLA/AO	SLA/MAO
S_q (μm)	0.198 ± 0.032	0.729 ± 0.070	0.544 ± 0.106	0.632 ± 0.056	0.825 ± 0.148

Surface hydrophilicity, which is one of the key factors promoting cellular response, was confirmed by monitoring the contact angle (Figure 5). The contact angle was assessed at particular intervals because wettability is a time-dependent feature after initial wetting of the surface [54]. The contact angle, which represents the degree of hydrophilicity, tends to decrease as the roughness increased for surface-treated Ti specimens. Among the various specimens, the HA specimen exhibited a similar level of hydrophilicity to that of the SLA/MAO specimen because HA present on the surface is known to produce good hydrophilicity. Significant hydrophilicity difference was not observed in SLA/AO compared to Ti. Nonetheless, SLA/AO specimens, whose roughness was greater than that of SLA specimens, showed greater hydrophilicity than that of SLA specimens. Previous research demonstrated that the application of SLA reduces the wettability of Ti [55], whereas the application of MAO increases the wettability of Ti [54]. This tendency was clearly confirmed by Figure 5b, which indicates that hydrophilicity was considerably greater for HA and SLA/MAO specimens than for other specimens in every case. This finding suggests that HA and SLA/MAO specimens would display the most desirable cellular behavior. In particular, contact angles on HA and SLA/MAO specimens decreased with time due to the ease of penetration of the micro-sized irregularities [54]. Various studies reported that an increase in wettability enhances cellular properties [56,57].

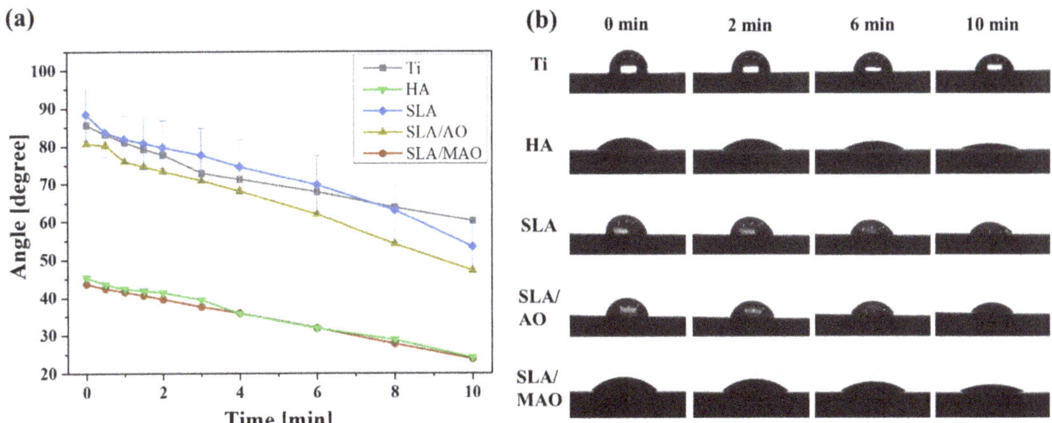

Figure 5. Measurement of wettability of surface-modified Ti through contact angle with water: (**a**) graph of the change in contact angle of each specimen over time and (**b**) monitored morphology of water drops on each specimen following dwelling time.

As a method to compare the stability of the coating, adhesion strength was measured using a pull-out test, which is suitable for measuring the required force for the removal of the coating from the substrate [58–60]. As suggested in Figure 6a–c, the coating layer could be clearly distinguished from the detached parts following the contours (red dashed line). Compared to HA and SLA/AO specimens, SLA/MAO specimens showed a more rugged boundary morphology along the remaining coating layer because of the much thicker TiO_2 layer due to MAO, as illustrated in Figure 2. Measured adhesion strengths were noted in Figure 6d: Adhesion in electrically treated specimens (SLA/AO and SLA/MAO) was significantly tighter than that in HA specimens. Unlike AO and MAO, HA is based mainly on physical treatment that induces collisions between accelerated HA particles and the target substrate. This difference led to a weaker bonding strength for HA than that for SLA/AO or SLA/MAO. The adhesion strength of the HA layer to the Ti substrate was comparable to that of a Polyetherimide(PEI) coating on Ti, since both of coating layers were physically entangled with the Ti substrate.

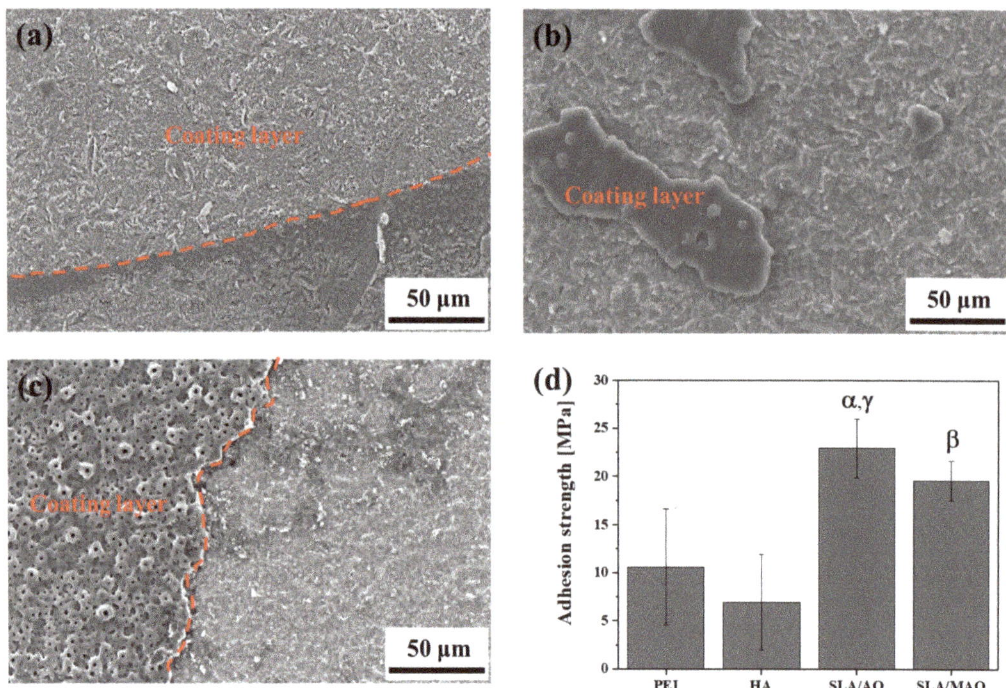

Figure 6. Surface morphology after detachment of coating layer of (**a**) HA, (**b**) SLA/AO, and (**c**) SLA/MAO. (**d**) quantitatively measured adhesion strength (α denotes $p < 0.05$ compared to PEI, β denotes $p < 0.05$ and γ denotes $p < 0.01$ compared to HA).

Results of in vitro tests of biocompatibility are illustrated in Figure 7. SEM images of the adhered cells after 3 h of incubation are presented in Figure 7a–e. The attached cells exhibited well-spread shapes on all of the specimens, demonstrating that the selected techniques were not cytotoxic to the cells. Since Ti itself is one of the metals with good cell affinity, an excellent degree of cell adhesion was found. Among the surface-treated specimens, the cells were found to be spreading more filopodia toward the surfaces in HA and SLA/MAO specimens, which showed high extent of surface roughness. In the case of the HA specimen, both the chemical similarity to the bone and the physically roughened surface complexly affected cellular behavior. When SLA, SLA/AO, and SLA/MAO were compared, the extension of cells was high in SLA/AO and SLA/MAO, driven by the increased roughness of the additional TiO_2 layer. Cell characteristics are generally known to be affected by the roughness and topography of the surface structure [29,61]. According to several previous studies regarding surface treatment on Ti, cell affinity is improved when surface area and surface roughness are enhanced by a nanotube structure or micron-sized roughness such as SLA/AO or SLA/MAO [62,63]. By contrast, the inter-treatment pattern in degree of proliferation was different from the pattern in cell attachment. Figure 7f illustrates cell viability as assessed by the MTS assay after 3 days or 5 days of culturing on each specimen. Measured cell viability at 3 days of culturing time presented a similar degree of cell proliferation, indicating that adhered cells multiplied comparably in the early stage. However, the obtained cell viability at 5 days of culturing had a different trend. Except for HA, cells proliferated considerably in all of the experimental groups; surface-treated Ti specimens provided an environment for cell growth preferable to that of pristine Ti specimens. In HA-Ti specimens, the instability of the coating layer, as inferred from the lower adhesion strength, may have resulted in an insecure state, slowing the proliferation of cells. According to Brodie et al., even if the same coating layer exists on the surface, the poor stability of the coating layer adversely affects cell proliferation [64]. The other specimens, whose coating layers were stable, differed meaningfully only in surface structure and composition, and all had a greater degree of proliferation than pristine Ti.

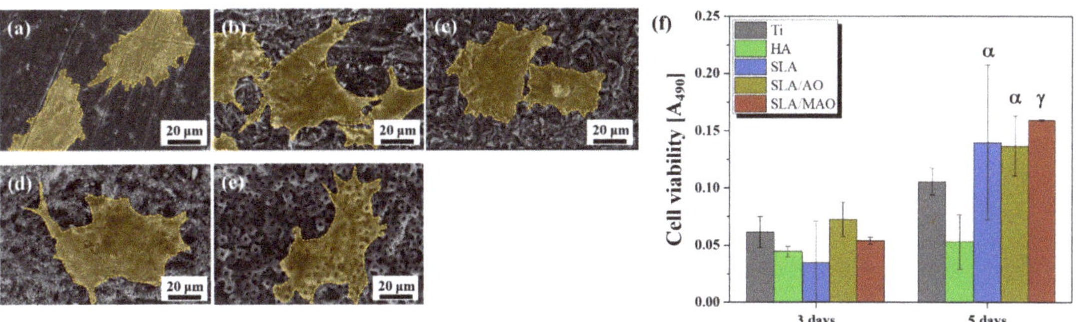

Figure 7. SEM images of attached pre-osteoblast cells on (**a**) Ti, (**b**) HA, (**c**) SLA, (**d**) SLA/AO, (**e**) SLA/MAO, and (**f**) measured degree of cell proliferation (α denotes $p < 0.05$ and γ denotes $p < 0.01$ compared to HA at 5 days).

4. Conclusions

In this study, four of the most prevalent surface treating strategies for Ti were systematically compared: HA-blasting, SLA, SLA/AO, and SLA/MAO. After surface treatment, each of the specimens exhibited a substantially roughened surface structure due to the additional coating layer or carving process. Elemental components were analyzed by EDS and XRD, confirming that Ti, HA, and TiO_2 layers were present, in agreement with previous research. Surface-treated Ti substrates exhibited higher hydrophilicity than that of pristine Ti substrates as a result of increased roughness and porosity. Furthermore, adhesion strength tests indicated that the HA coating layer was less stable than the TiO_2 layer formed through AO and MAO. In vitro test results demonstrated the superiority of electrical treatments (SLA/AO and SLA/MAO) in enhancing cellular attachment and proliferation.

Author Contributions: Conceptualization, T.-S.J. and H.-D.J.; methodology, H.L., D.K., C.-B.Y., G.H.; validation, J.K., H.-D.J.; investigation, J.K., H.L., C.-B.Y., G.H.; resources, H.L., D.K., H.-D.J.; data curation, J.K., H.L., H.-D.J.; writing—original draft preparation, J.K., H.L.; writing—review and editing, J.K., H.L., H.-D.J.; supervision, H.-E.K., H.-D.J.; project administration, H.-D.J.; funding acquisition, H.-E.K., H.-D.J.; All authors have read and agreed to the published version of the manuscript.

Funding: This work was supported by The Catholic University of Korea, Research Fund, 2020 and the Basic Science Research Program [No. 2020R1F1A1072103] through the National Research Foundation of Korea funded by the Korea government (MSIT).

Institutional Review Board Statement: Not applicable.

Informed Consent Statement: Not applicable.

Data Availability Statement: Not included.

Conflicts of Interest: The authors declare no conflict of interest.

References

1. Paital, S.R.; Dahotre, N.B. Calcium phosphate coatings for bio-implant applications: Materials, performance factors, and methodologies. *Mater. Sci. Eng. R Rep.* **2009**, *66*, 1–70. [CrossRef]
2. An, S.-H.; Matsumoto, T.; Miyajima, H.; Nakahira, A.; Kim, K.-H.; Imazato, S. Porous zirconia/hydroxyapatite scaffolds for bone reconstruction. *Dent. Mater.* **2012**, *28*, 1221–1231. [CrossRef]

3. Park, S.-B.; Lih, E.; Park, K.-S.; Joung, Y.K.; Han, D.K. Biopolymer-based functional composites for medical applications. *Prog. Polym. Sci.* **2017**, *68*, 77–105. [CrossRef]
4. Tan, A.C.W.; Polo-Cambronell, B.J.; Provaggi, E.; Ardila-Suárez, C.; Ramirez-Caballero, G.E.; Baldovino-Medrano, V.G.; Kalaskar, D.M. Design and development of low cost polyurethane biopolymer based on castor oil and glycerol for biomedical applications. *Biopolymers* **2017**, *109*, e23078. [CrossRef]
5. Jaafar, A.; Hecker, C.; Árki, P.; Joseph, Y. Sol-Gel Derived Hydroxyapatite Coatings for Titanium Implants: A Review. *Bioengineering* **2020**, *7*, 127. [CrossRef]
6. Hashemi, R. Failure Analysis of Biometals. *Metals* **2020**, *10*, 662. [CrossRef]
7. Kato, K.; Yamamoto, A.; Ochiai, S.; Wada, M.; Daigo, Y.; Kita, K.; Omori, K. Cytocompatibility and mechanical properties of novel porous 316L stainless steel. *Mater. Sci. Eng. C* **2013**, *33*, 2736–2743. [CrossRef]
8. Shah, F.A.; Omar, O.; Suska, F.; Snis, A.; Matic, A.; Emanuelsson, L.; Norlindh, B.; Lausmaa, J.; Thomsen, P.; Palmquist, A. Long-term osseointegration of 3D printed CoCr constructs with an interconnected open-pore architecture prepared by electron beam melting. *Acta Biomater.* **2016**, *36*, 296–309. [CrossRef]
9. Li, N.; Zheng, Y. Novel Magnesium Alloys Developed for Biomedical Application: A Review. *J. Mater. Sci. Technol.* **2013**, *29*, 489–502. [CrossRef]
10. Takemoto, M.; Fujibayashi, S.; Neo, M.; Suzuki, J.; Kokubo, T.; Nakamura, T. Mechanical properties and osteoconductivity of porous bioactive titanium. *Biomaterials* **2005**, *26*, 6014–6023. [CrossRef]
11. Carlsson, L.; Röstlund, T.; Albrektsson, B.; Albrektsson, T.; Brånemark, P.-I. Osseointegration of titanium implants. *Acta Orthop. Scand.* **1986**, *57*, 285–289. [CrossRef]
12. Shah, F.A.; Trobos, M.; Thomsen, P.; Palmquist, A. Commercially pure titanium (cp-Ti) versus titanium alloy (Ti6Al4V) materials as bone anchored implants—Is one truly better than the other? *Mater. Sci. Eng. C* **2016**, *62*, 960–966. [CrossRef]
13. Prachar, P.; Bartáková, S.; Vanek, J. The titanium PV I endosteal implant from beta-titanium alloy Ti 38Nb 6Ta. *Biomed. Pap.* **2015**, *159*, 503–507. [CrossRef]
14. Niinomi, M.; Nakai, M.; Hieda, J. Development of new metallic alloys for biomedical applications. *Acta Biomater.* **2012**, *8*, 3888–3903. [CrossRef]
15. Wally, Z.J.; Van Grunsven, W.; Claeyssens, F.; Goodall, R.; Reilly, G.C. Porous Titanium for Dental Implant Applications. *Metals* **2015**, *5*, 1902–1920. [CrossRef]
16. Aziz-Kerrzo, M.; Conroy, K.G.; Fenelon, A.M.; Farrell, S.T.; Breslin, C.B. Electrochemical studies on the stability and corrosion resistance of titanium-based implant materials. *Biofabrication* **2001**, *22*, 1531–1539. [CrossRef]
17. Cheng, A.; Humayun, A.; Cohen, D.J.; Boyan, B.D.; Schwartz, Z. Additively manufactured 3D porous Ti-6Al-4V constructs mimic trabecular bone structure and regulate osteoblast proliferation, differentiation and local factor production in a porosity and surface roughness dependent manner. *Biofabrication* **2014**, *6*, 045007. [CrossRef]
18. Nune, K.C.; Kumar, A.; Misra, R.D.K.; Li, S.J.; Hao, Y.L.; Yang, R. Osteoblast functions in functionally graded Ti-6Al-4 V mesh structures. *J. Biomater. Appl.* **2015**, *30*, 1182–1204. [CrossRef]
19. He, Y.; Zhang, Y.; Meng, Z.; Jiang, Y.; Zhou, R. Microstructure evolution, mechanical properties and enhanced bioactivity of Ti-Nb-Zr based biocomposite by bioactive calcium pyrophosphate. *J. Alloy. Compd.* **2017**, *720*, 567–581. [CrossRef]
20. Ozan, S.; Lin, J.; Li, Y.; Ipek, R.; Wen, C. Development of Ti–Nb–Zr alloys with high elastic admissible strain for temporary orthopedic devices. *Acta Biomater.* **2015**, *20*, 176–187. [CrossRef]
21. Cordeiro, J.M.; Nagay, B.E.; Ribeiro, A.L.R.; da Cruz, N.C.; Rangel, E.C.; Fais, L.M.; Vaz, L.G.; Barão, V.A. Functionalization of an experimental Ti-Nb-Zr-Ta alloy with a biomimetic coating produced by plasma electrolytic oxidation. *J. Alloy. Compd.* **2019**, *770*, 1038–1048. [CrossRef]
22. Stráský, J.; Harcuba, P.; Václavová, K.; Horváth, K.; Landa, M.; Srba, O.; Janeček, M. Increasing strength of a biomedical Ti-Nb-Ta-Zr alloy by alloying with Fe, Si and O. *J. Mech. Behav. Biomed. Mater.* **2017**, *71*, 329–336. [CrossRef] [PubMed]
23. Kolli, R.P.; Devaraj, A. A Review of Metastable Beta Titanium Alloys. *Metals* **2018**, *8*, 506. [CrossRef]
24. Chen, L.-Y.; Cui, Y.-W.; Zhang, L.-C. Recent Development in Beta Titanium Alloys for Biomedical Applications. *Metals* **2020**, *10*, 1139. [CrossRef]
25. Lee, H.; Jung, H.-D.; Kang, M.-H.; Song, J.; Kim, H.-E.; Jang, T.-S. Effect of HF/HNO3-treatment on the porous structure and cell penetrability of titanium (Ti) scaffold. *Mater. Des.* **2018**, *145*, 65–73. [CrossRef]
26. Trueba, P.; Beltrán, A.M.; Bayo, J.M.; Rodríguez-Ortiz, J.A.; Larios, D.F.; Alonso, E.; Dunand, D.C.; Torres, Y. Porous Titanium Cylinders Obtained by the Freeze-Casting Technique: Influence of Process Parameters on Porosity and Mechanical Behavior. *Metals* **2020**, *10*, 188. [CrossRef]
27. Kim, S.W.; Jung, H.-D.; Kang, M.-H.; Kim, H.-E.; Koh, Y.-H.; Estrin, Y. Fabrication of porous titanium scaffold with controlled porous structure and net-shape using magnesium as spacer. *Mater. Sci. Eng. C* **2013**, *33*, 2808–2815. [CrossRef]
28. Jung, H.-D.; Yook, S.-W.; Jang, T.-S.; Li, Y.; Kim, H.-E.; Koh, Y.-H. Dynamic freeze casting for the production of porous titanium (Ti) scaffolds. *Mater. Sci. Eng. C* **2013**, *33*, 59–63. [CrossRef]
29. Lukaszewska-Kuska, M.; Wirstlein, P.; Majchrowski, R.; Dorocka-Bobkowska, B. Osteoblastic cell behaviour on modified titanium surfaces. *Micron* **2018**, *105*, 55–63. [CrossRef]
30. Li, T.; Gulati, K.; Wang, N.; Zhang, Z.; Ivanovski, S. Understanding and augmenting the stability of therapeutic nanotubes on anodized titanium implants. *Mater. Sci. Eng. C* **2018**, *88*, 182–195. [CrossRef]

31. Wang, Y.; Yu, H.; Chen, C.; Zhao, Z. Review of the biocompatibility of micro-arc oxidation coated titanium alloys. *Mater. Des.* **2015**, *85*, 640–652. [CrossRef]
32. Jang, T.-S.; Jung, H.-D.; Kim, S.; Moon, B.-S.; Baek, J.; Park, C.; Song, J.; Kim, H.-E. Multiscale porous titanium surfaces via a two-step etching process for improved mechanical and biological performance. *Biomed. Mater.* **2017**, *12*, 025008. [CrossRef]
33. Lee, H.; Jang, T.-S.; Song, J.; Kim, H.-E.; Jung, H.-D. Multi-scale porous Ti_6Al_4V scaffolds with enhanced strength and biocompatibility formed via dynamic freeze-casting coupled with micro-arc oxidation. *Mater. Lett.* **2016**, *185*, 21–24. [CrossRef]
34. Han, C.-M.; Kim, H.-E.; Koh, Y.-H. Creation of hierarchical micro/nano-porous TiO_2 surface layer onto Ti implants for improved biocompatibility. *Surf. Coat. Technol.* **2014**, *251*, 226–231. [CrossRef]
35. Civantos, A.; Giner, M.; Trueba, P.; Lascano, S.; Montoya-García, M.-J.; Arévalo, C.; Vázquez, M. Ángeles; Allain, J.P.; Torres, Y. In Vitro Bone Cell Behavior on Porous Titanium Samples: Influence of Porosity by Loose Sintering and Space Holder Techniques. *Metals* **2020**, *10*, 696. [CrossRef]
36. Jung, H.-D.; Yook, S.-W.; Han, C.-M.; Jang, T.-S.; Kim, H.-E.; Koh, Y.-H.; Estrin, Y. Highly aligned porous Ti scaffold coated with bone morphogenetic protein-loaded silica/chitosan hybrid for enhanced bone regeneration. *J. Biomed. Mater. Res. Part B Appl. Biomater.* **2013**, *102*, 913–921. [CrossRef] [PubMed]
37. Jung, H.-D.; Jang, T.-S.; Wang, L.; Kim, H.-E.; Koh, Y.-H.; Song, J. Novel strategy for mechanically tunable and bioactive metal implants. *Biofabrication* **2015**, *37*, 49–61. [CrossRef]
38. Kim, S.; Park, C.; Cheon, K.-H.; Jung, H.-D.; Song, J.; Kim, H.-E.; Jang, T.-S. Antibacterial and bioactive properties of stabilized silver on titanium with a nanostructured surface for dental applications. *Appl. Surf. Sci.* **2018**, *451*, 232–240. [CrossRef]
39. Heimann, R.B. Osseoconductive and Corrosion-Inhibiting Plasma-Sprayed Calcium Phosphate Coatings for Metallic Medical Implants. *Metals* **2017**, *7*, 468. [CrossRef]
40. Kim, H.; Choi, S.-H.; Ryu, J.-J.; Koh, S.-Y.; Park, J.-H.; Lee, I.-S. The biocompatibility of SLA-treated titanium implants. *Biomed. Mater.* **2008**, *3*, 025011. [CrossRef]
41. Ferguson, S.; Broggini, N.; Wieland, M.; De Wild, M.; Rupp, F.; Geis-Gerstorfer, J.; Cochran, D.; Buser, D. Biomechanical evaluation of the interfacial strength of a chemically modified sandblasted and acid-etched titanium surface. *J. Biomed. Mater. Res. Part A* **2006**, *78*, 291–297. [CrossRef] [PubMed]
42. Huang, C.-F.; Chiang, H.-J.; Lin, H.-J.; Hosseinkhani, H.; Ou, K.-L.; Peng, P.-W. Comparison of Cell Response and Surface Characteristics on Titanium Implant with SLA and SLAffinity Functionalization. *J. Electrochem. Soc.* **2014**, *161*, G15–G20. [CrossRef]
43. He, W.; Yin, X.; Xie, L.; Liu, Z.; Li, J.; Zou, S.; Chen, J. Enhancing osseointegration of titanium implants through large-grit sandblasting combined with micro-arc oxidation surface modification. *J. Mater. Sci. Mater. Med.* **2019**, *30*, 73. [CrossRef] [PubMed]
44. Kim, S.; Park, C.; Moon, B.-S.; Kim, H.-E.; Jang, T.-S. Enhancement of osseointegration by direct coating of rhBMP-2 on target-ion induced plasma sputtering treated SLA surface for dental application. *J. Biomater. Appl.* **2016**, *31*, 807–818. [CrossRef]
45. Li, L.-H.; Kong, Y.-M.; Kim, H.-W.; Kim, Y.-W.; Kim, H.-E.; Heo, S.-J.; Koak, J.-Y. Improved biological performance of Ti implants due to surface modification by micro-arc oxidation. *Biofabrication* **2004**, *25*, 2867–2875. [CrossRef]
46. Tsutsumi, Y.; Niinomi, M.; Nakai, M.; Shimabukuro, M.; Ashida, M.; Chen, P.; Doi, H.; Hanawa, T. Electrochemical Surface Treatment of a β-titanium Alloy to Realize an Antibacterial Property and Bioactivity. *Metals* **2016**, *6*, 76. [CrossRef]
47. Wei, D.; Zhou, Y.; Wang, Y.; Jia, D. Characteristic of microarc oxidized coatings on titanium alloy formed in electrolytes containing chelate complex and nano-HA. *Appl. Surf. Sci.* **2007**, *253*, 5045–5050. [CrossRef]
48. Li, L.-H.; Kim, H.-W.; Lee, S.-H.; Kong, Y.-M.; Kim, H.-E. Biocompatibility of titanium implants modified by microarc oxidation and hydroxyapatite coating. *J. Biomed. Mater. Res. Part A* **2005**, *73*, 48–54. [CrossRef]
49. Wu, L.; Wen, C.; Zhang, G.; Liu, J.; Ma, K. Influence of anodizing time on morphology, structure and tribological properties of composite anodic films on titanium alloy. *Vacuum* **2017**, *140*, 176–184. [CrossRef]
50. Kuromoto, N.K.; Simão, R.A.; Soares, G.A. Titanium oxide films produced on commercially pure titanium by anodic oxidation with different voltages. *Mater. Charact.* **2007**, *58*, 114–121. [CrossRef]
51. Wang, H.-Y.; Zhu, R.-F.; Lu, Y.-P.; Xiao, G.-Y.; He, K.; Yuan, Y.; Ma, X.-N.; Li, Y. Effect of sandblasting intensity on microstructures and properties of pure titanium micro-arc oxidation coatings in an optimized composite technique. *Appl. Surf. Sci.* **2014**, *292*, 204–212. [CrossRef]
52. Zhan, X.; Li, S.; Cui, Y.; Tao, A.; Wang, C.; Li, H.; Linzhang, L.; Yu, H.; Jiang, J.; Li, C. Comparison of the osteoblastic activity of low elastic modulus Ti-24Nb-4Zr-8Sn alloy and pure titanium modified by physical and chemical methods. *Mater. Sci. Eng. C* **2020**, *113*, 111018. [CrossRef] [PubMed]
53. Graziano, A.; D'Aquino, R.; Angelis, M.G.C.-D.; De Francesco, F.; Giordano, A.; Laino, G.; Piattelli, A.; Traini, T.; De Rosa, A.; Papaccio, G. Scaffold's surface geometry significantly affects human stem cell bone tissue engineering. *J. Cell. Physiol.* **2007**, *214*, 166–172. [CrossRef] [PubMed]
54. Kulkarni, M.; Patil-Sen, Y.; Junkar, I.; Kulkarni, C.V.; Lorenzetti, M.; Iglič, A. Wettability studies of topologically distinct titanium surfaces. *Colloids Surf. B Biointerfaces* **2015**, *129*, 47–53. [CrossRef] [PubMed]
55. Hotchkiss, K.M.; Reddy, G.B.; Hyzy, S.L.; Schwartz, Z.; Boyan, B.D.; Olivares-Navarrete, R. Titanium surface characteristics, including topography and wettability, alter macrophage activation. *Acta Biomater.* **2016**, *31*, 425–434. [CrossRef]

56. Gittens, R.A.; Olivares-Navarrete, R.; Cheng, A.; Anderson, D.M.; McLachlan, T.; Stephan, I.; Geis-Gerstorfer, J.; Sandhage, K.H.; Fedorov, A.G.; Rupp, F.; et al. The roles of titanium surface micro/nanotopography and wettability on the differential response of human osteoblast lineage cells. *Acta Biomater.* **2013**, *9*, 6268–6277. [CrossRef]
57. Rosales-Leal, J.; Rodríguez-Valverde, M.; Mazzaglia, G.; Ramón-Torregrosa, P.; Díaz-Rodríguez, L.; García-Martínez, O.; Vallecillo-Capilla, M.; Ruiz, C.; Cabrerizo-Vílchez, M. Effect of roughness, wettability and morphology of engineered titanium surfaces on osteoblast-like cell adhesion. *Colloids Surf. A Physicochem. Eng. Asp.* **2010**, *365*, 222–229. [CrossRef]
58. Kim, S.-B.; Jo, J.-H.; Lee, S.-M.; Kim, H.-E.; Shin, K.-H.; Koh, Y.-H. Use of a poly(ether imide) coating to improve corrosion resistance and biocompatibility of magnesium (Mg) implant for orthopedic applications. *J. Biomed. Mater. Res. Part A* **2013**, *101*, 1708–1715. [CrossRef]
59. Cheon, K.-H.; Park, C.; Kang, M.-H.; Kang, I.-G.; Lee, M.-K.; Lee, H.; Kim, H.-E.; Jung, H.-D.; Jang, T.-S. Construction of tantalum/poly(ether imide) coatings on magnesium implants with both corrosion protection and osseointegration properties. *Bioact. Mater.* **2021**, *6*, 1189–1200. [CrossRef]
60. Kang, M.-H.; Jang, T.-S.; Jung, H.-D.; Kim, S.-M.; Kim, H.-E.; Koh, Y.-H.; Song, J. Poly(ether imide)-silica hybrid coatings for tunable corrosion behavior and improved biocompatibility of magnesium implants. *Biomed. Mater.* **2016**, *11*, 035003. [CrossRef]
61. İzmir, M.; Ercan, B. Anodization of titanium alloys for orthopedic applications. *Front. Chem. Sci. Eng.* **2019**, *13*, 28–45. [CrossRef]
62. Yu, W.Q.; Zhang, Y.L.; Jiang, X.Q.; Zhang, F.Q. In vitro behavior of MC3T3-E1 preosteoblast with different annealing temperature titania nanotubes. *Oral Dis.* **2010**, *16*, 624–630. [CrossRef] [PubMed]
63. Oh, S.; Daraio, C.; Chen, L.-H.; Pisanic, T.R.; Fiñones, R.R.; Jin, S. Significantly accelerated osteoblast cell growth on aligned TiO2 nanotubes. *J. Biomed. Mater. Res. Part A* **2006**, *78*, 97–103. [CrossRef] [PubMed]
64. Brodie, J.C.; Goldie, E.; Connel, G.; Merry, J.; Grant, M.H. Osteoblast interactions with calcium phosphate ceramics modified by coating with type I collagen. *J. Biomed. Mater. Res. Part A* **2005**, *73*, 409–421. [CrossRef] [PubMed]

Article

Discovery of New Ti-Based Alloys Aimed at Avoiding/Minimizing Formation of α'' and ω-Phase Using CALPHAD and Artificial Intelligence

Rajesh Jha and George S. Dulikravich *

Multidisciplinary Analysis, Inverse Design, Robust Optimization and Control (MAIDROC) Laboratory, Department of Mechanical and Materials Engineering, Florida International University, Miami, FL 33174, USA; rjha001@fiu.edu

* Correspondence: dulikrav@fiu.edu; Tel.: +1-954-554-0368

Abstract: In this work, we studied a Ti-Nb-Zr-Sn system for exploring novel composition and temperatures that will be helpful in maximizing the stability of β phase while minimizing the formation of α'' and ω-phase. The Ti-Nb-Zr-Sn system is free of toxic elements. This system was studied under the framework of CALculation of PHAse Diagram (CALPHAD) approach for determining the stability of various phases. These data were analyzed through artificial intelligence (AI) algorithms. Deep learning artificial neural network (DLANN) models were developed for various phases as a function of alloy composition and temperature. Software was written in Python programming language and DLANN models were developed utilizing TensorFlow/Keras libraries. DLANN models were used to predict various phases for new compositions and temperatures and provided a more complete dataset. This dataset was further analyzed through the concept of self-organizing maps (SOM) for determining correlations between phase stability of various phases, chemical composition, and temperature. Through this study, we determined candidate alloy compositions and temperatures that will be helpful in avoiding/minimizing formation of α'' and ω-phase in a Ti-Zr-Nb-Sn system. This approach can be utilized in other systems such as ω-free shape memory alloys. DLANN models can even be used on a common Android mobile phone.

Keywords: Ti-based biomaterials; biocompatibility; toxicity; β-phase; ω-phase; CALPHAD; artificial intelligence; deep learning artificial neural network (DLANN); self-organizing maps (SOM)

Citation: Jha, R.; Dulikravich, G.S. Discovery of New Ti-Based Alloys Aimed at Avoiding/Minimizing Formation of α'' and ω-Phase Using CALPHAD and Artificial Intelligence. *Metals* **2021**, *11*, 15. https://dx.doi.org/10.3390/met11010015

Received: 28 November 2020
Accepted: 19 December 2020
Published: 24 December 2020

1. Introduction

Titanium-based alloys have been widely accepted for biomedical applications due to comparatively superior biocompatibility and anti-corrosion properties [1–5]. Efforts are being made to explore new alloys that contain elements that are not toxic, as well as develop alloys with Young's modulus comparable to human bone to avoid stress shielding [2]. Young's modulus (YM) of common implant materials varies between 100–230 GPa, which is significantly higher when compared with that of bone, which is between 10 and 40 GPa [1]. This difference in YM results in non-uniform distribution of stress in the implant materials and the surrounding bone structure. This can result in the failure of an implant [1]. Titanium alloys containing β-phase as the predominant phase are known to have lower values of Young's modulus [2–9]. Thermodynamically, α and β are stable phases, while α'' and ω-phase are metastable phases [2]. Composition and processing of alloys play an important role in determining the concentrations of various phases, which directly affect mechanical properties of these alloys [6–9].

During processing or heat-treatment of Ti-based alloys, they are subjected to cooling from an elevated temperature at various cooling rates [10–14]. The cooling rate can affect the stability of β-phase and it can transform into α, α'' or ω-phase, while ω-phase can also form isothermally during ageing [12]. Among these phases, ω-phase possesses the highest modulus value. Precipitation of α and ω-phase can result in an increase in the Young's modulus of the alloy [12,13]. Regarding α''-phase, it is a desired phase in shape memory alloys [12]. In Ti-based biomaterials, precipitation of α, α'' or ω-phase can lead to degradation of mechanical properties, specifically Young's modulus [2,10–14]. In titanium alloys, experiments show that ω-phase can be stabilized and has been observed at cryogenic temperatures [15]. From first-principle calculations, researchers have demonstrated the scope of development of ω-phase free Ti-Ta-X shape memory alloys [16].

In this work, we chose the Ti-Nb-Zr-Sn system, which is free of toxic chemical elements [17] such as aluminum [18] and vanadium [19]. Niobium is a strong β-phase stabilizer [9]. In the presence of Nb, Zr also becomes a strong β-phase stabilizer [1]. Additionally, Nb and Zr are biocompatible and demonstrate excellent resistance to corrosion [20]. Sn is also biocompatible and has been used as an alloying element in small amounts [21]. The addition of Sn leads to a decrease in elastic modulus of Ti-based biomaterials [22]. Sn addition can help in suppressing precipitation of α'' and ω-phase for lower and higher cooling rates, respectively [22]. One of the aims is to optimize the alloy composition so that an optimum amount of Sn addition can prove to be beneficial in the lowering of the elastic modulus and in the suppression of precipitation of α'' and ω-phase [22].

The Ti-Nb-Zr-Sn system was studied under the framework of CALculation of PHAse Diagram (CALPHAD) approach through Thermo-Calc software [23]. It was used for determining concentrations of various stable and metastable phases for a large set of compositions and temperatures, and thus generating a dataset suitable enough for applying artificial intelligence (AI) algorithms, including deep learning [24] and self-organizing maps (SOM) [25]. These data were then used for developing deep learning artificial neural network (DLANN) models for various phases as a function of alloy composition and temperature. Software for developing the DLANN model was written in Python programming language using TensorFlow [26] and Keras [27] libraries. DLANN models were used as a predictive tool for predicting the concentrations of metastable phases for new compositions and temperatures. This resulting dataset was then analyzed by the concept of self-organizing maps (SOM) [25] for determining various patterns and correlations within the dataset among alloying elements, temperatures, and stable and metastable phases.

Through this work, we were able to identify the composition and temperature regimes that will provide ω-phase free Ti-based alloys with a minimal amount of α'' phase. We have reported chemical compositions of several candidate alloys along with temperatures that will be helpful in achieving ω-phase free Ti-based alloys (Ti-Nb-Zr-Sn system). Our research team has significant experience in designing alloys by application of artificial intelligence (AI) algorithms on data generated from experiments and data generated under the framework of the CALPHAD approach. We have successfully designed titanium alloys [28], aluminum alloys [29,30], hard magnets (AlNiCo) [31,32], soft magnets (FINEMET type) [33,34], and Ni-based superalloys [35]. Thus, we propose this computational design approach, which can be easily adopted in other alloy systems and can help in developing ω-phase free Ti-based biomaterials with improved mechanical properties.

2. Materials and Methods

As mentioned, we chose the Ti-Nb-Zr-Sn system for this work. Bounds for concentrations for each of the four alloying elements were defined on the basis of available literature [1–21]. Bounds for temperature include cryogenic temperature [15] for determining the amount of ω-phase in the presence and absence of thermodynamically stable phases. Variable bounds for the chemical composition of the Ti-Nb-Zr-Sn system are presented in Table 1. Chemical compositions are in atomic %, while the temperature is in degrees Kelvin. Phase stability calculations for approximately 3000 candidate alloys were performed for random values of concentrations and temperatures within theranges reported in Table 1. There is no correlation between the minimum and maximum values of composition with the minimum and maximum values of temperature reported in Table 1. Combinations of alloy composition and temperature are randomly generated in five-dimensional (5-D) space for achieving a uniform distribution of support points in 5-D space shown in Table 1. Uniform distribution of support points is helpful in improving the accuracy of prediction for models developed through AI algorithms.

Table 1. Concentration bounds for alloying elements (atomic %) for the Ti-Nb-Zr-Sn system and temperature range (K) chosen for study.

	Nb	Zr	Sn	Ti	Temp. (K)
Minimum	0.03	0.02	0.01	58.9	50.0
Maximum	31.6	9.95	4.95	97.7	1526.0

2.1. Identification of Stable and Metastable Phases

We used commercial software Thermo-Calc [23,36] along with thermodynamic database TCTI2 [37] for titanium alloys for performing phase stability calculations. In the Ti-Nb-Zr-Sn system, α and β are stable phases, while α″ and ω-phase are metastable phases. Regarding the crystal structure, α phase has a hexagonal close packed (HCP) structure, while β phase is body centered cubic (BCC). In the TCTI2 database [37], HCP_A3 is the notation for α phase. Regarding β phase, it exists in two forms: BCC_B2 and BCC_B2#2. Metastable phase α″ is denoted by ALTI3_D019, while metastable phase ω is denoted by OMEGA in TCTI2 database [37]. These notations were also mentioned in our previous publication on titanium-based alloys [28], which is featured on Thermo-Calc's website [38]. We have provided our previous publication on titanium alloys as a reference since the titanium database was recently launched by Thermo-Calc in 2018. There are only a few references on thorough studies of various stable and metastable phases in titanium alloys using Thermo-Calc. While performing phase stability calculations, metastable phases are unstable in the presence of stable phases and thus are absent from phase stability data. However, these metastable phases have been observed in the microstructure obtained after performing experiments [39]. Among α, β, α″ and ω-phase, α and β are thermodynamically stable, while α″ and ω phases are metastable [2].

For multi-component systems, several equilibria exist for a given composition and temperature. The CALPHAD approach works on the principles of Gibbs energy minimization. Phases that have the lowest Gibbs energy of formation values are thermodynamically stable, and other phases are deemed unstable or metastable. Through the CALPHAD approach, only the thermodynamically stable phases will be shown on the phase diagram. Metastable phases cannot be shown on the phase diagram, nor is there any estimate of the amount (mole) of these phases. Metastable phases can be preferentially stabilized by suppressing/removing a set of stable phases while performing phase stability calculations [29,33,34,40]. Once a set of thermodynamically stable phases are suppressed/removed, we are left with the remaining phases of that system. In the absence of thermodynamically stable phases, few metastable phases become stable as now these phases possess the lowest Gibbs energy of formation values among the remaining phases. Through this preferential stabilization, one can obtain an estimate of the concentrations of metastable phases that can exist under equilibrium conditions for a given composition and temperature. A brief understanding of the formation of metastable phases for a given system can be helpful while designing experiments including heat treatment protocols. We have published articles on stabilizing metastable phases in aluminum alloys [29] and soft magnetic FINEMET-type alloys [33,34]. In soft magnetic alloys, we even performed heat treatment simulations for studying the nucleation and growth of metastable phases and some of the results were experimentally verified using advanced diagnostic tools such as atom probe tomography [34]. Thus, in this work, we were able to generate a large dataset comprised of about 3000 randomly selected candidate alloy compositions and temperatures along with various stable and metastable phases.

2.2. Deep Learning Artificial Neural Network (DLANN) Model

We developed DLANN models within the framework of deep learning [24] These models were coded in Python programming language and used TensorFlow [26] and Keras [27] libraries for developing this code. For visualization, we used Tensorboard [41]. DLANN models were developed separately for each of the phases. A separate dataset was prepared for each phase and the data were scaled between 0 and 1. These scaled data were randomly divided into training and testing sets where 33% data were assigned to the testing set while the remaining 67% were included in the training set. DLANN architecture includes 4 hidden layers. The number of neurons in the initial hidden layer was fixed at 50, 60, 70, 80, 90, and 100, while the number of neurons in the other three hidden layers was fixed at 100, 120, 140, 160, 180 and 200 neurons. In short, the number of neurons in the initial layer was half of the number of neurons in each of the other three layers. Activation function was Rectified Linear Unit (ReLU) and optimizer was "Adam" which stands for Adaptive Moment Estimation, while number of epochs was fixed at 100. Tensorboard was used for visualizing the DLANN performance [41]. One of the criteria for the selection of a model was its performance on validation set, which was determined through error metrics such as mean squared error (MSE) and mean absolute error (MAE). One has to be careful while relying on error metrics such as MAE and MSE alone as we are dealing with ANN models that are susceptible to "overfitting". Although we are dealing with a large dataset, there are still many "missing" points. Therefore, we used statistical terms for guidance, while we gave priority to physical metallurgy of titanium alloys for DLANN model selection. These DLANN models can be used on a personal computer and even on an Android mobile phone to predict various metastable and stable phases, and thus provide us with a dataset with additional support points for determining various patterns and correlations within this dataset.

2.3. Self Organizing Maps (SOM)

Data obtained from CALPHAD-based calculations and predicted through DLANN models were further studied by the concept of self-organizing maps (SOM) [25,28,31,42]. SOM maps are known for preserving topology of the data, which is helpful in determining various correlations in the dataset among concentrations of alloying element, temperature, and stability of various stable and metastable phases. Regarding prediction via SOM, one must be careful while drawing conclusions as SOM values over a hexagonally-shaped cell are the average value of candidates positioned at the vertices of the hexagonally-shaped cells. Thus, for prediction we used DLANN models, while for determining correlations over a large dataset of about 3000 candidate alloys, we considered SOM. Through this study, we identified several candidate alloys that are free of ω-phase. For SOM analysis, we used a commercial software ESTECO-modeFRONTIER, version 4.5, Trieste, Italy [42].

2.4. Computational Infrastructure

2.4.1. CALPHAD-Based Work

Thermo-Calc version 2019b [23,37] was installed on a desktop computer in a computer lab. The operating system was Windows 10, Core i7 processor (CPU) with 16 GB RAM. Phase transformation calculation time varied between 20 and 30 min.

2.4.2. Artificial Intelligence-Based Work

AI-based work was performed on a laptop. The operating system was Windows 10, Core i7 processor (CPU) with 32 GB RAM. DLANN model development took about 20 to 30 min. Once the model was developed, prediction was completed in a few seconds. DLANN models were also used on an Android phone with 6 GB RAM and octa-core processor (CPU). Prediction time is a few seconds on the Android phone. SOM model development took about 20–30 min for each case.

3. Results

3.1. Stability of Stable and Metastable Phases

From CALPHAD-based calculations, one can observe the OMEGA (ω) phase at cryogenic temperatures as low as 50 K [15]. At these temperatures, phase stability calculations provided a high amount of the OMEGA (ω) phase (above 0.8 mole fraction). With a rise in temperature, the OMEGA (ω) phase decreased and reached zero even below room temperature. In order to stabilize the OMEGA (ω) phase, few phases including HCP_A3 (α), BCC_B2 (β) and ALTI3_D019 (α″) were suppressed/removed while performing phase stability calculations. This stabilized the OMEGA (ω) phase, and we were able to stabilize the OMEGA (ω) phase at higher temperatures through the CALPHAD approach. From experiments, it has been confirmed that the OMEGA (ω) phase is present in the same amount after processing/heat treatment [8–15]. Thus, it was important to stabilize the OMEGA (ω) phase for better understanding of its formation and stability over a large temperature range.

Figure 1 shows the relative comparison of the occurrence of the OMEGA (ω) phase over a large range of temperature (0–1500 K). In Figure 1, the entire temperature range was divided into five parts. About 3000 candidate alloy compositions were analyzed in this temperature range. The number of cases was recorded for which the OMEGA (ω) phase was observed in each of these temperature ranges. Regarding legends, "α and β stable" means HCP_A3 (α) and BCC_B2 (β) were included in the phase stability calculations and both phases were stable. Legend "α″ only" means that in this case, both HCP_A3 (α) and BCC_B2 (β) were removed while performing equilibrium calculations for stabilizing ALTI3_D019 (α″) phase. Legend "ω only" means that HCP_A3 (α), BCC_B2 (β) and ALTI3_D019 (α″) phases were removed while performing phase stability calculations for stabilizing the OMEGA (ω) phase.

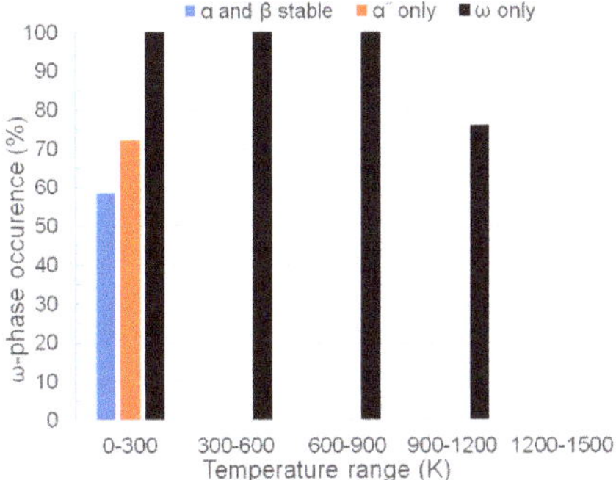

Figure 1. Occurrence (%) of the OMEGA (ω) phase for different temperature ranges for three separate equilibrium calculations.

From Figure 1, we can see that the OMEGA (ω) phase can be stabilized at elevated temperatures through the CALPHAD approach as observed through experiments [8–15]. In order to stabilize the OMEGA (ω) phase at elevated temperatures through the CALPHAD approach, one needs to remove HCP_A3 (α), BCC_B2 (β) and ALTI3_D019 (α″) phases along with a few other phases while performing phase stability calculations. Through the CALPHAD approach, a user needs to perform separate calculations each time they need to analyze a particular composition or temperature for determining metastable phases. This approach is time consuming as a user needs to have access to the computer on which CALPHAD-based software is installed.

Next, we move forward to application of AI algorithms on phase stability data generated through the CALPHAD approach. AI algorithms will be helpful in developing accurate predictive models that can capture trends and patterns within a large dataset.

3.2. DLANN Model

As mentioned before, DLANN models were selected on the basis of physical metallurgy of titanium alloys as well as on error metrics. DLANN architecture and error metrics (MSE and MAE) over the validation set are listed in Table 2. From Table 2, one can notice that values of MSE are acceptable, but values of MAE are a bit high. The amount of phase varied between zero and one for each of the stable and metastable phases included in this work, while MAE varies between 0.01783 to 0.03574. Thus, MAE for this work is between approximately 1.8% and 3.6% of the maximum amount of any phase. We have mentioned before that 67% of data were assigned to the training set and 33% of data were included in the testing or validation set. Thus, there is room for improvement in prediction accuracy (error metrics) by increasing the amount of data in the training set. However, while working on accuracy, we must be careful as ANN models are susceptible to "overfitting". Thus, based upon physical metallurgy of titanium alloys, error metrics in the present case and our own experience in handling such problems, we selected the models listed in Table 2 for further analysis.

Table 2. Performance metrics for deep learning artificial neural network (DLANN) models for various phases for the Ti-Nb-Zr-Sn system.

Phase	DLANN Architecture	Error Metrics (Validation Set)	
		Mean Absolute Error (MAE)	Mean Squared Error (MSE)
ALTI3_D019(α″)	50-100-100-100	0.03286	0.00549
BCC_B2 (β)	80-160-160-160	0.03182	0.00608
BCC_B2#2	90-180-180-180	0.03574	0.00916
HCP_A3(α)	90-180-180-180	0.01783	0.00135
OMEGA (ω)	70-140-140-140	0.01922	0.00248

DLANN models were used as a predictive tool and can be used on a computer and even on an Android device. As mentioned before, metastable phases are absent in the presence of stable phases while performing phase stability calculations under the framework of the CALPHAD approach [29,33,34,39]. We used the alloy composition and temperatures included in the dataset obtained from initial calculations containing only stable phases and then predicted metastable phases for these alloy compositions and temperatures through DLANN models. Thus, DLANN models were used to obtain an improved dataset for further analysis through SOM.

3.3. Self-Organizing Maps (SOM)

SOM analysis [25,28,31,42] was performed on the data obtained through the CALPHAD approach and DLANN models. From CALPHAD and DLANN analysis, we have a matrix of 3000 rows and nine columns. Here, rows are 3000 candidate alloys. Columns are alloy compositions (Ti, Nb, Zr, Sn), temperature and the phases BCC_B2, HCP_A3, ALTI3_D019, and the OMEGA phase. Thus, we included all the design variables and the objectives. Calculations were performed in batch mode, where all the designs are introduced to the SOM algorithm with value of X unit set at 15 and Y unit assigned a value of 18 [28,31]. Thus, there are 270 map units on a SOM map. Each map unit is in the form of a hexagonal cell and candidate alloys are positioned at the vertices of the hexagonal units. The 3000 candidate alloys along with temperature and concentration of phase values are arranged over 270 units on the SOM maps on the basis of algorithm setting. Other parameters were optimized so that SOM maps are able to capture trends in the dataset [28,31,42].

In this work, we used a commercial software ESTECO-modeFRONTIER for SOM analysis [42]. This software provides a user with two types of error values: quantization error and topological error. Quantization refers to the ability of the SOM algorithm to learn from data distribution. As mentioned, about 3000 candidate alloy compositions and temperatures and amounts of stable and metastable phases are presented in batch mode. These 3000 candidates are positioned at the vertices of 270 hexagonal unit cells on SOM maps. SOM analysis provides these candidates with new prototype positions on the SOM map. Quantization error is an estimate of the average distance between the initial position of a candidate and its prototype position assigned through SOM analysis.

The SOM algorithm is known for preserving the topology of the dataset. As mentioned, there are 270 hexagonal map units and candidates are arranged on each of these units. Through topology error, the algorithm checks for the relative position of a candidate with respect to candidates positioned in adjacent hexagonal map units. Thus, initially all the candidates are positioned on the SOM maps and as per SOM algorithm settings, all of these candidates are assigned new prototype positions. Through topology error, the SOM algorithm determines the relative distance between initial and prototype positions of candidates in the neighboring hexagonal units. This way, all the candidates positioned on the SOM maps are checked.

The SOM model was chosen on the basis of error metrics of a model and capability of a model to mimic trends shown in the literature for Ti-based biomaterials. Physical metallurgy of Ti-based alloys was given a priority while error metrics acted as a guiding tool. SOM error metrics have been reported in Table 3. Here, we can observe that model error for SOM is quite low. Hence, we moved ahead with analyzing the SOM maps for understanding patterns within the dataset.

Table 3. Self-organizing maps (SOM) error metrics for the Ti-Nb-Zr-Sn system.

Quantization Error	Topological Error
0.064	0.028

Figure 2 shows the SOM component plot for the Ti-Nb-Zr-Sn system. For SOM analysis, BCC_B2#2 phase was not included as there were too many missing points and also due to the fact that it is another form of same BCC_B2 phase included in the TCTI2 [37] database. From Figure 2, we can observe that BCC_B2(β) and HCP_A3(α) are positioned together. Components positioned together are correlated in SOM maps. From physical metallurgy of titanium alloys, we know that titanium alloys in practice are either predominantly α or β, or a mixture of both in different proportions [2,28]. Thus, the stability of α and β phases is correlated from a metallurgical point of view. The SOM algorithm was able to determine correlations that can be verified from reported works on titanium alloys, even though the SOM algorithm is an unsupervised machine learning approach and does not work on the principle of Gibbs energy minimization [25,28,31].

ALTI3_D019 (α″) is close to HCP_A3 (α) and can be correlated. The OMEGA (ω) phase is far enough from other cells so we cannot confirm that it is correlated with the other components. Temperature is below BCC_B2 (β) and HCP_A3 (α) and close to Sn and Zr. Temperature is not close enough to these components and we cannot provide a concluding remark on the correlation between temperature and other components. Elements Ti and Nb are clustered together similar to Zr and Sn. The OMEGA (ω) phase is close to Ti and Nb, but not close enough to point towards strong correlation. Thus, SOM analysis provided us with vital information on various strong and weak correlations among alloying elements, stable and metastable phases, and temperature for the Ti-Nb-Zr-Sn system. Now, we will proceed further to analyze each of these components.

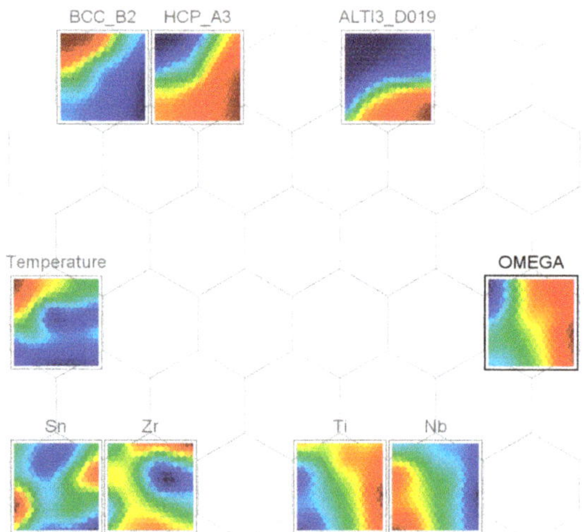

Figure 2. SOM components plot for the Ti-Nb-Zr-Sn system.

Figure 3 shows the SOM maps for HCP_A3 (α), BCC_B2 (β), ALTI3_D019 (α″) and OMEGA (ω) phases along with chemical concentrations of Nb, Zr and Sn and temperature. From Figure 3, one can observe that for temperature the lowest value on the color bar is 645 K and the highest value is 1232 K, while in Table 1, the range of temperature was between 50 K and 1526 K. The reason for this is that we have analyzed about 3000 candidate alloys through SOM. As mentioned before, each candidate is placed on the vertices of hexagonal cells on SOM maps. The SOM algorithm is used for pattern recognition in small to large and often multi-dimensional datasets. In SOM maps, various regions are marked on the bases of average values of candidate alloys placed on the vertices of a hexagonal cell. Thus, a region marked 645 K in the figure consists of six candidate alloys for which the average temperature is about 645 K.

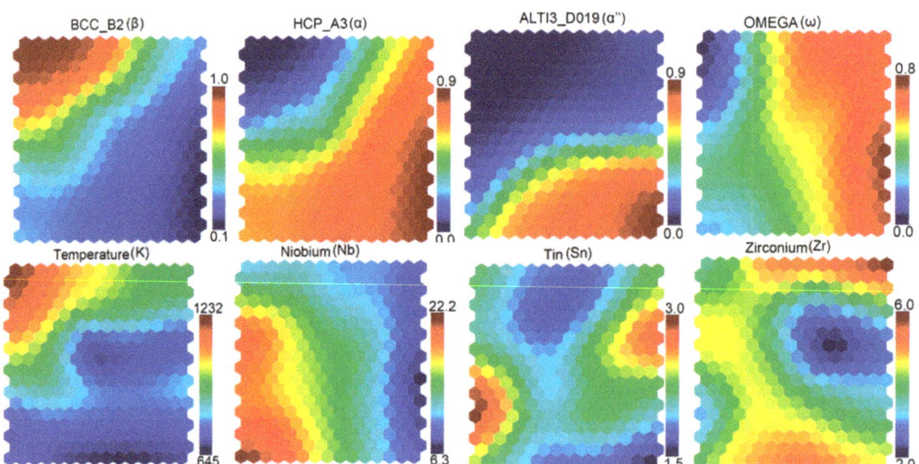

Figure 3. SOM plot showing chemical concentrations, temperatures and resulting stable and metastable phases for the Ti-Nb-Zr-Sn system.

From literature, we know that HCP_A3 (α) is stable at lower temperatures and BCC_B2 (β) is stable at higher temperatures [2,28]. In Figure 3, we can observe that same pattern for HCP_A3 (α) and BCC_B2 (β) phases. At higher temperatures, one can fully stabilize BCC_B2 (β) phase, while suppressing formation of HCP_A3 (α), ALTI3_D019 (α″) and OMEGA (ω) phase. With respect to composition, one needs to design compositions in a way that Nb is between average to low value, Sn is below average value and Zr is average and below average. Here, the average value refers to the color bar for the compositions in Figure 3.

Figure 4 shows the distribution for titanium and temperature along with BCC_B2 (β), HCP_A3 (α), ALTI3_D019 (α″) and OMEGA (ω) phase. From Figure 4, one can observe that a user must maintain titanium at average composition as shown through color bar in the figure. At the average titanium composition, along with elevated temperature, a user can design compositions that will be predominantly the BCC_B2 (β) phase and these candidates are expected to be free from the HCP_A3 (α), ALTI3_D019 (α″) and OMEGA (ω) phase.

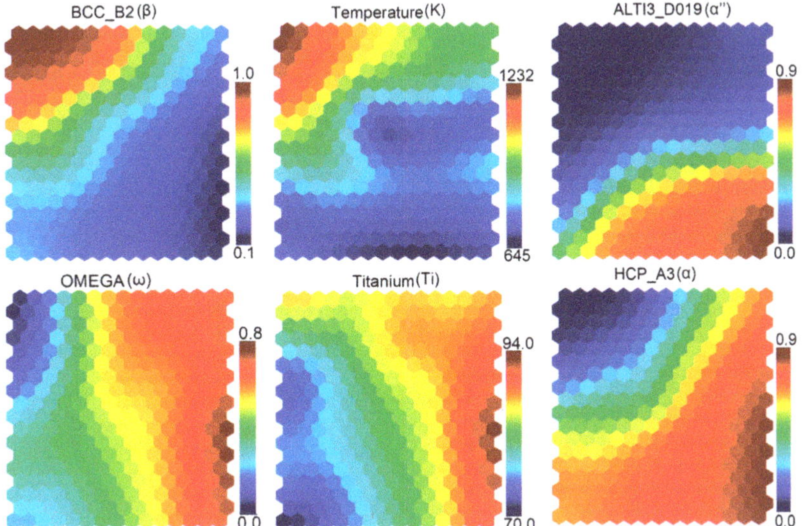

Figure 4. SOM plot showing composition (Ti), temperature and resulting stable and metastable phases for the Ti-Nb-Zr-Sn system.

From Figures 3 and 4, one can observe that OMEGA (ω) predicted through the DLANN model is stable for a wide range of temperatures and compositions. Figure 1 shows a similar trend of occurrence of the OMEGA (ω) phase over a wide temperature range. Figure 1 was plotted using data from CALPHAD-based calculations, where a user needs to perform calculations separately for stabilizing metastable phases. Through AI algorithms, all of this can be achieved at the same instant. AI-based predictions can be performed on a normal computer for free as we have developed our code in Python language, which is free.

From this work, five candidate alloy compositions and temperatures were identified (Table 4). These alloys are expected to have a fully stabilized BCC_B2 (β) phase and to be free from other phases such as HCP_A3 (α), ALTI3_D019 (α″), and OMEGA (ω) phase. For these select alloys, the amount of OMEGA (ω) phase obtained through phase stability calculations, stabilizing OMEGA (ω) phase and value of OMEGA (ω) phase predicted through DLANN models and SOM maps were all zero.

Table 4. Candidate alloys predicted through CALculation of PHAse Diagram (CALPHAD), DLANN models and SOM approach, with zero concentrations of HCP_A3 (α), ALTI3_D019 (α″), and OMEGA (ω) phases.

Alloy No.	Ti (Mole %)	Nb (Mole %)	Zr (Mole %)	Sn (Mole %)	Temp. (K)
1	63.11244	29.85438	5.11341	1.91977	641.7
2	65.56413	27.26226	6.32562	0.84798	807.825
3	65.44534	27.15556	6.62609	0.77301	955.644
4	64.15603	26.81113	7.31843	1.71441	989.739
5	73.28298	23.13284	1.39528	2.1889	1024.94

4. Discussion

This research problem had the main goal of determining the compositions and temperatures for Ti-Nb-Zr-Sn alloys, which will provide an alloy that is predominantly BCC_B2 (β) phase and free from other phases such as HCP_A3 (α), ALTI3_D019 (α″), and OMEGA (ω) phase. In this work, this task was accomplished through combined CALPHAD and artificial intelligence (AI).

We identified one publication [2] on thermodynamic modeling on Ti-based biomaterials, which can be compared with our current work. In that work [2], the author performed first-principle calculations along with thermodynamic modeling within the framework of the CALPHAD approach for predicting metastable phases in the Ti-Nb-Zr-Sn-Ta system [2]. The author listed as his future work that he will work on the development of models based on first-principle calculations for predicting the α″ and ω phase [2] and indicated plans to study the effect of Sn addition in larger amounts in order to study its effect on the stability of the β phase [2]. Another work [16] based on shape memory alloys can also be compared with the present work. In their work [16], the authors performed first-principle calculations for developing ω-phase free Ti-Ta-X systems. Both of these references are thorough works and have included results from first-principle calculations [2,16]. Density functional theory (DFT) or first-principle calculations are computationally expensive, and a user needs to have access to supercomputers for performing DFT-based study. Additionally, one of these works [2] was performed in 2017 when Thermo-Calc did not have a commercially available database for Ti-based alloys. In the last few years, there has been significant development in improving the database of Ti-based alloys [23,36–39]. The Ti-based alloy database now includes several new elements, which means several new equilibriums [37]. Many new models have been included for predicting various stable and metastable phases [37]. Thus, in the current work, it was possible to address a few of the limitations mentioned in these references [2].

In this work, we used Thermo-Calc [23,36] along with the TCTI2 database [23,36–39]. Our objective was to accelerate the process of discovery of new alloy compositions for Ti-based biomaterials and temperatures at which the β phase is fully stabilized. Thus, we relied on existing CALPHAD-based models and generated data for stability of various stable and metastable phases. Thereafter, we chose to develop models for various stable and metastable phases through the application of artificial intelligence algorithms.

Notice that no work was presented on improving the models for α″ and ω-phase through first-principle calculations, as this was not within the scope of the present work. The purpose was to develop models that can be used for predicting the concentrations of stable and metastable phases in a few seconds. Consequently, DLANN models developed in this work can be used on a personal computer and even on a normal Android phone.

The SOM algorithm was further helpful in determining various correlations among chemical compositions, temperatures, and concentrations of stable and metastable phases. Determining these correlations were mentioned in the future work of one of the articles that dealt with thermodynamic modeling [2]. The current work demonstrates that it is possible to efficiently predict a few candidate alloys that are expected to meet requirements regarding the stability of β phase.

Future Work

Regarding future work, we plan to work on the following topics:

- Expanding the scope of study regarding working with new alloying elements that are biocompatible and non-toxic [2,13,43] by applying CALPHAD and AI approaches for exploring new compositions and temperatures of new alloys.
- Develop predictive models for Young's modulus of new proposed alloys through the CALPHAD approach and AI algorithms.
- Study kinetics of precipitation of various stable and metastable phases within the framework of the CALPHAD approach and work with solidification simulation to have a better understanding of precipitation of various stable and metastable phases for different cooling rates. Thereafter, study precipitation kinetics of nucleation and growth of various phases.
    - This study will be helpful for understanding micro-segregation, especially for cast prosthetics [13,44–46]. Studies have shown that during solidification, it is difficult to avoid composition variation in the inter-dendritic region due to solute entrapment, which thus makes the casting composition non-homogeneous [44]. Micro-segregation can be controlled by properly choosing the cooling rate [13,44]. Thus, solidification simulation will be helpful in understanding the temperature regimes where a certain desired or undesired phase is stable [44]. This way, one should be able to design a cooling rate that is fast enough to avoid ageing in the temperature regimes where undesired phases are unstable.
    - Heat-treatment simulations are equally important [46]. Some of these alloys are subjected to ageing at a defined temperature for a prolonged time (several hours). Through heat treatment simulations, one can obtain an estimate of the grain size and volume fractions of a desired phase and observe its growth over time. Grain size and volume fraction affect the Young's modulus of an alloy, so this study is important.
- Simulate microstructure evolution, micro-segregation, composition variation in the inter-dendritic regions [47–49] under the framework of the CALPHAD and phase field approach [47–49].
    - The phase field approach is a popular approach for simulating microstructure evolution. A user can get insights required for the understanding of the solidification process and can study the growth of dendrites and composition variation in inter-dendritic regions, which is important for addressing micro-segregation [47–49]. The CALPHAD approach will be used for providing vital information on thermodynamics and kinetics to the phase-field models especially regarding the sequence of precipitation of a phase as well as stability of various phases [49]. The CALPHAD approach also provides the grain size, and this information can be used to calibrate the phase field model [49].
- Design new manufacturing protocols with special emphasis on additive manufacturing [50–56].
    - Manufacturing of parts via additive manufacturing is a viable method, especially for users who need custom made implants [50–57]. The additive manufacturing route is also helpful in developing implants with lower Young's modulus and improved biocompatibility [51].
    - Several modes of designing new parts through additive manufacturing exist, such as selective laser beam, electron beam, etc. [50–52,56]. All of these methods have advantages and limitations [50–52,56]. Optimization of operation param-

eters plays a vital role in achieving targeted properties of a prosthetic/implant manufactured by additive manufacturing [50,53].
    - CALPHAD, and the phase field approach have been used for studying microstructure evolution for additively manufactured parts [47–49]. AI algorithms have been used to study data and develop inexpensive predictive models within the framework of additive manufacturing [57]. We plan to work on these topics.
- Finally, the most important characteristic of an implant is its biocompatibility, and osteointegration [55,58–62]. Several coatings have been developed and there is always room for improvement [55,58–62]. We plan to use AI-based tools to understand these coatings and possibly design new coatings with enhanced biocompatibility and osteointegration.

5. Conclusions

In this work, we proposed a novel, computationally efficient approach for accelerating the discovery of new compositions and process parameters for Ti-based biomaterials that will help in achieving fully stabilized β-phase, which is required for improving multiple desired properties of existing biomaterials. This work utilizes the information reported in literature to generate data for various stable and metastable phases under the framework of the CALPHAD approach. AI algorithms were used to accelerate the discovery of new compositions and temperatures.

This work can be summarized as follows:

- Data for various stable and metastable phases were generated for about 3000 composition and temperature values of a Ti-Nb-Zr-Sn system through the commercial software Thermo-Calc and TCTI database for titanium alloys.
- Phase stability data were used for developing deep learning artificial neural network (DLANN) models for various phases as a function of alloy composition and temperature. DLANN models were used to predict the concentrations of phases for new compositions and temperatures. DLANN models can be used on a personal computer and even on an Android phone.
- The SOM algorithm was used to determine correlations among alloying elements, temperature, and various stable and metastable phases.
- Finally, we predicted compositions of five select alloys that are expected to meet our expectations regarding the phase stability of β phase.

Author Contributions: Conceptualization, R.J. and G.S.D.; methodology, R.J.; software, R.J.; validation, R.J.; formal analysis, R.J.; investigation, R.J.; resources, R.J. and G.S.D.; data curation, R.J.; writing—original draft preparation, R.J.; writing—review and editing, R.J. and G.S.D.; visualization, R.J.; supervision, G.S.D.; project administration, G.S.D.; funding acquisition, G.S.D. All authors have read and agreed to the published version of the manuscript.

Funding: This work was supported by the College of Engineering and Computing at Florida International University and by NASA HQ University Leadership Initiative (ULI) program under federal award number NNX17AJ96A titled "Adaptive Aerostructures for Revolutionary Supersonic Transportation" managed by Texas A&M University.

Acknowledgments: Authors would like to express their sincere gratitude to Professor Cristian V. Ciobanu from Colorado School of Mines, Golden, Colorado, USA, and Professor Carlo Poloni from ESTECO Co., Trieste, Italy for providing access to some of the commercial software used in this paper. Reviewers' comments and suggestions helped the readability of this paper and are highly appreciated.

Conflicts of Interest: The authors declare no conflict of interest. The funders had no role in the design of the study; in the collection, analyses, or interpretation of data; in the writing of the manuscript, or in the decision to publish the results.

References

1. Long, M.; Rack, H. Titanium alloys in total joint replacement—A materials science perspective. *Biomaterials* **1998**, *19*, 1621–1639. [CrossRef]
2. Marker, C. Development of a Knowledge Base of Ti-Alloys from First-Principles and Thermodynamic Modeling. Ph.D. Thesis, The Pennsylvania State University, State College, PA, USA, August 2017. Available online: https://www.proquest.com/docview/1988756108 (accessed on 30 October 2019).
3. Jung, H.D.; Jang, T.S.; Wang, L.; Kim, H.E.; Koh, Y.H.; Song, J. Novel strategy for mechanically tunable and bioactive metal implants. *Biomaterials* **2015**, *37*, 49–61. [CrossRef] [PubMed]
4. Lee, H.; Jang, T.S.; Song, J.; Kim, H.E.; Jung, H.D. Multi-scale porous Ti6Al4V scaffolds with enhanced strength and biocompatibility formed via dynamic freeze-casting coupled with micro-arc oxidation. *Mater. Lett.* **2016**, *185*, 21–24. [CrossRef]
5. Jang, T.S.; Kim, D.; Han, G.; Yoon, C.B.; Jung, H.D. Powder based additive manufacturing for biomedical application of titanium and its alloys: A review. *Biomed. Eng. Lett.* **2020**, *10*, 505–516. [CrossRef]
6. Kolli, R.P.; Devaraj, A. A Review of Metastable Beta Titanium Alloys. *Metals* **2018**, *8*, 506. [CrossRef]
7. Mohammed, M.T.; Khan, Z.A.; Siddiquee, A.N. Beta titanium alloys: The lowest elastic modulus for biomedical applications: A review. *Int. J. Chem. Nucl. Metall. Mater. Eng.* **2014**, *8*, 726–731.
8. Soundararajan, S.R.; Vishnu, J.; Manivasagam, G.; Muktinutalapati, N.R. Processing of Beta Titanium Alloys for Aerospace and Biomedical Applications. In *Titanium Alloys—Novel Aspects of Their Processing*; IntechOpen: London, UK, 2018.
9. Magdalen, H.C.; Baghi, A.D.; Ghomashchi, R.; Xiao, W.; Oskouei, R.H. Effect of niobium content on the microstructure and Young's modulus of Ti-xNb-7Zr alloys for medical implants. *J. Mech. Behav. Biomed.* **2019**, *99*, 78–85.
10. Banerjee, S.; Tewari, R.; Dey, G.K. Omega phase transformation—Morphologies and mechanisms. *Int. J. Mater. Res.* **2006**, *97*, 963–977. [CrossRef]
11. Mantri, S.; Choudhuri, D.; Behera, A.; Hendrickson, M.; Alam, T.; Banerjee, R. Role of isothermal omega phase precipitation on the mechanical behavior of a Ti-Mo-Al-Nb alloy. *Mater. Sci. Eng. A* **2019**, *767*, 138397. [CrossRef]
12. Li, Y.; Yang, C.; Zhao, H.; Qu, S.; Li, X.; Li, Y. New Developments of Ti-Based Alloys for Biomedical Applications. *Materials* **2014**, *7*, 1709–1800. [CrossRef]
13. Manivasagam, G.; Singh, A.K.; Asokamani, R.; Gogia, A.K. Ti based biomaterials, the ultimate choice for orthopaedic implants—A review. *Prog. Mater. Sci.* **2009**, *54*, 397–425. [CrossRef]
14. Aleixo, G.T.; Afonso, C.; Coelho, A.; Caram, R. Effects of Omega Phase on Elastic Modulus of Ti-Nb Alloys as a Function of Composition and Cooling Rate. *Solid State Phenom.* **2008**, *138*, 393–398. [CrossRef]
15. De Fontaine, D.; Paton, N.; Williams, J. The omega phase transformation in titanium alloys as an example of displacement controlled reactions. *Acta Met.* **1971**, *19*, 1153–1162. [CrossRef]
16. Ferrari, A.; Paulsen, A.; Langenkämper, D.; Piorunek, D.; Somsen, C.; Frenzel, J.; Rogal, J.; Eggeler, G.; Drautz, R. Discovery of ω-free high-temperature Ti-Ta-X shape memory alloys from first-principles calculations. *Phys. Rev. Mater.* **2019**, *3*, 103605. [CrossRef]
17. Ito, A.; Okazaki, Y.; Tateishi, T.; Ito, Y. In vitro biocompatibility, mechanical properties, and corrosion resistance of Ti-Zr-Nb-Ta-Pd and Ti-Sn-Nb-Ta-Pd alloys. *J. Biomed. Mater. Res.* **1995**, *29*, 893–899. [CrossRef]
18. Boyce, B.F.; McWilliams, S.; Mocan, M.Z.; Elder, H.Y.; Boyle, I.T.; Junor, B.J.R. Histological and electron microprobe studies of mineralization in aluminum related osteomalacia. *J. Clin. Pathol.* **1992**, *45*, 502–508. [CrossRef]
19. Domingo, J.L. Vanadium and Tungsten Derivatives as Antidiabetic Agents. *Biol. Trace Elem. Res.* **2002**, *88*, 97–112. [CrossRef]
20. Tane, M.; Akita, S.; Nakano, T.; Hagihara, K.; Umakoshi, Y.; Niinomi, M.; Nakajima, H. Peculiar elastic behavior of Ti–Nb–Ta–Zr single crystals. *Acta Mater.* **2008**, *56*, 2856–2863. [CrossRef]
21. Niinomi, M.; Nakai, M.; Hieda, J. Development of new metallic alloys for biomedical applications. *Acta Biomater.* **2012**, *8*, 3888–3903. [CrossRef]
22. Aleixo, G.T.; Lopes, E.; Contieri, R.; Cremasco, A.; Afonso, C.R.; Caram, R. Effects of Cooling Rate and Sn Addition on the Microstructure of Ti-Nb-Sn Alloys. *Solid State Phenom.* **2011**, *172*, 190–195. [CrossRef]
23. Andersson, J.-O.; Helander, T.; Höglund, L.; Shi, P.; Sundman, B. Thermo-Calc & DICTRA, computational tools for materials science. *Calphad* **2002**, *26*, 273–312. [CrossRef]
24. LeCun, Y.; Bengio, Y.; Hinton, G. Deep learning. *Nature* **2015**, *521*, 436–444. [CrossRef] [PubMed]
25. Kohonen, T. Self-organized formation of topologically correct feature maps. *Biol. Cybern.* **1982**, *43*, 59–69. [CrossRef]
26. TensorFlow. Available online: https://www.tensorflow.org/ (accessed on 30 May 2020).
27. Keras: The Python Deep Learning Library. Available online: https://keras.io/ (accessed on 30 May 2020).
28. Jha, R.; Dulikravich, G.S. Design of High Temperature Ti–Al–Cr–V Alloys for Maximum Thermodynamic Stability Using Self-Organizing Maps. *Metals* **2019**, *9*, 537. [CrossRef]
29. Jha, R.; Dulikravich, G.S. Solidification and heat treatment simulation for aluminum alloys with scandium addition through CALPHAD approach. *Comput. Mater. Sci.* **2020**, *182*, 109749. [CrossRef]
30. Jha, R.; Dulikravich, G.S. Determination of Composition and Temperature Regimes for Stabilizing Precipitation Hardening Phases in Aluminum Alloys with Scandium Addition: Combined CALPHAD—Deep Learning Approach. *Calphad* **2021**.
31. Jha, R.; Dulikravich, G.S.; Chakraborti, N.; Fan, M.; Schwartz, J.; Koch, C.C.; Colaco, M.J.; Poloni, C.; Egorov, I.N. Self-organizing maps for pattern recognition in design of alloys. *Mater. Manuf. Process.* **2017**, *32*, 1067–1074. [CrossRef]

32. Jha, R.; Dulikravich, G.S.; Colaco, M.J.; Fan, M.; Schwartz, J.; Koch, C.C. Magnetic Alloys Design Using Multi-Objective Optimization. In *Advanced Structured Materials*; Oechsner, A., Da Silva, L.M., Altenbac, H., Eds.; Springer: Berlin/Heidelberg, Germany, 2017; Volume 33, pp. 261–284, 978–981.
33. Jha, R.; Chakraborti, N.; Diercks, D.R.; Stebner, A.P.; Ciobanu, C.V. Combined machine learning and CALPHAD approach for discovering processing-structure relationships in soft magnetic alloys. *Comput. Mater. Sci.* **2018**, *150*, 202–211. [CrossRef]
34. Jha, R.; Diercks, D.R.; Chakraborti, N.; Stebner, A.P.; Ciobanu, C.V. Interfacial energy of copper clusters in Fe-Si-B-Nb-Cu alloys. *Scr. Mater.* **2019**, *162*, 331–334. [CrossRef]
35. Jha, R.; Pettersson, F.; Dulikravich, G.S.; Saxen, H.; Chakraboti, N. Evolutionary Design of Nickel-Based Superalloys Using Data-Driven Genetic Algorithms and Related Strategies. *Mater. Manuf. Process.* **2015**, *30*, 488–510. [CrossRef]
36. Thermo-Calc Software: Thermo-Calc Version 2019b. Available online: https://thermocalc.com/release-news/ (accessed on 30 October 2019).
37. Thermo-Calc Software: TCTI2: TCS Ti/TiAl-Based Alloys Database. Available online: https://thermocalc.com/products/databases/titanium-and-titanium-aluminide-based-alloys/ (accessed on 30 October 2019).
38. Thermo-Calc Software. Available online: https://thermocalc.com/solutions/solutions-by-material/titanium-and-titanium-aluminide/ (accessed on 30 October 2020).
39. Thermo-Calc Software TCAL5: TCS Aluminium-Based Alloys Database v.5. Available online: https://www.thermocalc.com/media/56675/TCAL5-1_extended_info.pdf (accessed on 30 August 2019).
40. Assadiki, A.; Esin, V.A.; Bruno, M.; Martinez, R. Stabilizing effect of alloying elements on metastable phases in cast aluminum alloys by CALPHAD calculations. *Comput. Mater. Sci.* **2018**, *145*, 1–7. [CrossRef]
41. TensorBoard: TensorFlow's Visualization Toolkit. Available online: https://www.tensorflow.org/tensorboard (accessed on 30 May 2020).
42. ESTECO-ModeFRONTIER. Available online: https://www.esteco.com/technology/analytics-visualization (accessed on 30 May 2020).
43. Liang, S.; Feng, X.; Yin, L.; Liu, X.Y.; Ma, M.; Liu, R. Development of a new β Ti alloy with low modulus and favorable plasticity for implant material. *Mater. Sci. Eng. C* **2016**, *61*, 338–343. [CrossRef] [PubMed]
44. Stefanescu, D.M.; Ruxanda, R. Solidification Structures of Titanium Alloys. In *Metallography and Microstructures*; ASM International: Geauga, OH, USA, 2004; Volume 9, pp. 116–126.
45. Raabe, D.; Sander, B.; Friák, M.; Ma, D.; Neugebauer, J. Theory-guided bottom-up design of β-titanium alloys as biomaterials based on first principles calculations: Theory and experiments. *Acta Mater.* **2007**, *55*, 4475–4487. [CrossRef]
46. Kuroda, P.B.; Quadros, F.D.F.; De Araújo, R.O.; Afonso, C.R.; Grandini, C. Effect of Thermomechanical Treatments on the Phases, Microstructure, Microhardness and Young's Modulus of Ti-25Ta-Zr Alloys. *Materials* **2019**, *12*, 3210. [CrossRef] [PubMed]
47. Sahoo, S.; Chou, K. Phase-field simulation of microstructure evolution of Ti–6Al–4V in electron beam additive manufacturing process. *Addit. Manuf.* **2016**, *9*, 14–24. [CrossRef]
48. Ding, L.; Sun, Z.; Liang, Z.; Li, F.; Xu, G.; Chang, H. Investigation on Ti-6Al-4V Microstructure Evolution in Selective Laser Melting. *Metals* **2019**, *9*, 1270. [CrossRef]
49. Ahluwalia, R.; Laskowski, R.; Ng, N.; Wong, M.; Quek, S.S.; Wu, D.T. Phase field simulation of α/βmicrostructure in titanium alloy welds. *Mater. Res. Express* **2020**, *7*, 046517. [CrossRef]
50. Attar, H.; Ehtemam-Haghighi, S.; Soro, N.; Kent, D.; Dargusch, M.S. Additive manufacturing of low-cost porous titanium-based composites for biomedical applications: Advantages, challenges and opinion for future development. *J. Alloys Compd.* **2020**, *827*, 154263. [CrossRef]
51. Trevisan, F.; Calignano, F.; Aversa, A.; Marchese, G.; Lombardi, M.; Biamino, S.; Ugues, D.; Manfredi, D. Additive manufacturing of titanium alloys in the biomedical field: Processes, properties and applications. *J. Appl. Biomater. Funct. Mater.* **2018**, *16*, 57–67. [CrossRef]
52. Jakubowicz, J. Special Issue: Ti-Based Biomaterials: Synthesis, Properties and Applications. *Materials* **2020**, *13*, 1696. [CrossRef]
53. Majumdar, T.; Eisenstein, N.; Frith, J.E.; Cox, S.; Birbilis, N. Additive Manufacturing of Titanium Alloys for Orthopedic Applications: A Materials Science Viewpoint. *Adv. Eng. Mater.* **2018**, *20*, 1800172. [CrossRef]
54. Wei, J.; Sun, H.; Zhang, D.; Gong, L.; Lin, J.; Wen, C. Influence of Heat Treatments on Microstructure and Mechanical Properties of Ti–26Nb Alloy Elaborated In Situ by Laser Additive Manufacturing with Ti and Nb Mixed Powder. *Materials* **2018**, *12*, 61. [CrossRef] [PubMed]
55. Imagawa, N.; Inoue, K.; Matsumoto, K.; Ochi, A.; Omori, M.; Yamamoto, K.; Nakajima, Y.; Kato-Kogoe, N.; Nakano, H.; Matsushita, T.; et al. Mechanical, Histological, and Scanning Electron Microscopy Study of the Effect of Mixed-Acid and Heat Treatment on Additive-Manufactured Titanium Plates on Bonding to the Bone Surface. *Materials* **2020**, *13*, 5104. [CrossRef] [PubMed]
56. Calignano, F.; Galati, M.; Iuliano, L.; Minetola, P. Design of Additively Manufactured Structures for Biomedical Applications: A Review of the Additive Manufacturing Processes Applied to the Biomedical Sector. *J. Health Eng.* **2019**, *2019*, 9748212. [CrossRef]
57. Wang, Z.; Liu, P.; Xiao, Y.; Cui, X.; Hu, Z.; Chen, L. A Data-Driven Approach for Process Optimization of Metallic Additive Manufacturing Under Uncertainty. *J. Manuf. Sci. Eng.* **2019**, *141*, 1–51. [CrossRef]
58. Sidambe, A.T. Biocompatibility of Advanced Manufactured Titanium Implants—A Review. *Materials* **2014**, *7*, 8168–8188. [CrossRef]

59. Kuroda, K.; Okido, M. Hydroxyapatite Coating of Titanium Implants Using Hydroprocessing and Evaluation of Their Osteoconductivity. *Bioinorg. Chem. Appl.* **2012**, *2012*, 1–7. [CrossRef]
60. Lu, R.-J.; Wang, X.; He, H.-X.; Ling-Ling, E.; Li, Y.; Zhang, G.-L.; Li, C.-J.; Ning, C.-Y.; Liu, H. Tantalum-incorporated hydroxyapatite coating on titanium implants: Its mechanical and in vitro osteogenic properties. *J. Mater. Sci. Mater. Med.* **2019**, *30*, 111. [CrossRef]
61. Kazimierczak, P.; Przekora, A. Osteoconductive and Osteoinductive Surface Modifications of Biomaterials for Bone Regeneration: A Concise Review. *Coatings* **2020**, *10*, 971. [CrossRef]
62. Jaafar, A.; Hecker, C.; Árki, P.; Joseph, Y. Sol-Gel Derived Hydroxyapatite Coatings for Titanium Implants: A Review. *Bioengineering* **2020**, *7*, 127. [CrossRef]

 metals

Article

Histomorphometric Analysis of Osseointegrated Grade V Titanium Mini Transitional Implants inEdentulous Mandible by Backscattered Scanning Electron Microscopy (BS-SEM)

Víctor Beltrán [1,2], Benjamín Weber [2], Ricardo Lillo [3], María-Cristina Manzanares [4], Cristina Sanzana [5], Nicolás Fuentes [3], Pablo Acuña-Mardones [1] and Ivan Valdivia-Gandur [6,7,*]

1 Clinical Investigation and Dental Innovation Center (CIDIC), Dental School and Center for Translational Medicine (CEMT-BIOREN), Universidad de La Frontera, Temuco 4811230, Chile; victor.beltran@ufrontera.cl (V.B.); pablo.acuna@ufrontera.cl (P.A.-M.)
2 Postgraduate Program in Oral Rehabilitation, Department of Integral Adults Dentistry, Dental School, Universidad de La Frontera, Temuco 4811230, Chile; benjamin.weber@ufrontera.cl
3 Postgraduate Program in Oral Implantology, Dental School, Universidad Mayor, Santiago 7510041, Chile; rlilloe@gmail.com (R.L.); nicolas.fuentes.f@gmail.com (N.F.)
4 Muscular and Skeletal Pathology Research, Human Anatomy and Embryology Unit, Universitat de Barcelona, 08193 Barcelona, Spain; mcmanzanares@ub.edu
5 Institute of Dental Sciences and Postgraduate Program in Oral Implantology, Facultad de Odontología, Universidad de Chile, Santiago 8380544, Chile; csanzana@ug.uchile.cl
6 Dentistry Department, University of Antofagasta, Avenida Angamos 601, Antofagasta 1270300, Chile
7 Biomedical Department, University of Antofagasta, Avenida Angamos 601, Antofagasta 1270300, Chile
* Correspondence: ivan.valdivia@uantof.cl; Tel.: +56-55-637018

Citation: Beltrán, V.; Weber, B.; Lillo, R.; Manzanares, M.; Sanzana, C.; Fuentes, N.; Acuña-Mardones, P.; Valdivia-Gandur, I. Histomorphometric Analysis of Osseointegrated Grade V Titanium Mini Transitional Implants in Edentulous Mandible by Backscattered Scanning Electron Microscopy (BS-SEM). *Metals* **2021**, *11*, 2. https://dx.doi.org/10.3390/met11010002

Received: 30 October 2020
Accepted: 27 November 2020
Published: 22 December 2020

Abstract: The purpose of this study is to assess the use of grade V titanium mini transitional implants (MTIs) immediately loaded by a temporary overdenture. For this, a histomorphometric analysis of the bone area fraction occupancy (BAFO) was performed by backscattered scanning electron microscopy (BS-SEM). Four female patients were submitted to surgery in which two MTIs were installed and immediately loaded with a temporary acrylic prosthesis. During the same surgery, two regular diameter implants were placed inside the bone and maintained without mechanical load. After 8 months, the MTIs were extracted using a trephine and processed for ultrastructural bone analysis by BS-SEM, and the regular-diameter implants were loaded with an overdenture device. A total of 243 BAFOs of MTIs were analyzed, of which 94 were mainly filled with cortical bone, while 149 were mainly filled with trabecular bone. Bone tissue analysis considering the total BAFOs with calcified tissues showed 72.13% lamellar bone, 26.04% woven bone, and 1.82% chondroid bone without significant differences between the samples. This study revealed that grade V titanium used in immediately loaded MTI was successfully osseointegrated by a mature and vascularized bone tissue as assessed from the BAFO.

Keywords: grade V titanium; mini transitional implants; narrow diameter implant; backscattered electrons

1. Introduction

One alternative to addressing the overly narrow residual ridges of the mandible is the use of dental mini implants (with a diameter below 3 mm) [1], which represent an important therapeutic option in cases in which reconstructive surgeries are contraindicated or as an alternative to other surgical procedures that present greater morbidity [2]. For these cases, the use of alloys with improved mechanical properties is desired for use in dental implants [3]. In this context, grade V titanium mini transitional implants (MTIs) can be a better alternative in these clinical cases because of their high mechanical resistance. These procedures allow the clinician to perform restoration and maintenance work in the vertical dimension with the use of mini implants in conjunction with implant surgical therapy, which is considered to be an effective method to provide the patient with an immediate and comfortable transitional appliance [4].

Grade V titanium alloy is widely used to fabricate medical and dental implants due to its superior physical, mechanical [5,6], and biological properties compared to other available biomedical metals and alloys. Specifically, its resistance to corrosion, excellent biocompatibility, lightweight, and superior tribological properties make this material far superior to other metals and alloys and highly useful for dental application [7,8]. Moreover, various authors have demonstrated its osseointegration [9–11]. Complications related to biocompatibility [12] or toxicity [13] of titanium alloys have been reported, including a mildly increased inflammatory response in direct contact to skeletal muscle [14]. Still, the evidence that includes human biopsies is scant.

Ultrastructural bone analysis has been performed by backscattered scanning electron microscopy (BS-SEM). This is an effective technique for studying the osseointegration of dental implants [15]; however, there is scarce evidence regarding the osseointegration of grade V titanium dental implants in humans when considering transitional implants immediately loaded and associated with overdenture. The purpose of this study is to assess the use of grade V titanium MTIs as immediately loaded implants for temporary overdenture, performing the analysis of the bone tissue close to the implant surface in human biopsies.

2. Materials and Methods

2.1. Patients and Surgical Procedure

The study was approved by the Ethics Committee of the University Mayor, Chile (Protocol N°246/006). Four female patients (60–68 years of age, health compatible with treatment) who were non-smokers, users of acrylic removable dentures over the completely edentulous mandible, and with horizontal atrophy of the anterior area were selected for treatment using an overdenture which was fixed temporally by MTIs. Prior to implant placement, all patients signed for informed consent regarding description of the placement and removal of the MTIs. The surgical protocol involved the simultaneous placement of submerged regular implants (4.0 mm Titamax EX or 4.3 mm Alvim Morse Taper) and MTIs for immediate loading in the edentulous lower mandible. The submerged regular implants that were not loaded were maintained and controlled for their osseointegration in the traditional way (6–8 months to implant osseointegration).

Each patient received two MTIs (2.9 mm in diameter and 12 mm long, Facility Neodent with Neoporos, Curitiba, Brazil) that were placed in the canine mandible region with O-rings for fixation of the overdentures. The MTIs were placed with a mandatory insertion torque of at least 35 Ncm and maximum 45 Ncm measured by a drill unit (W&H Implant Med). The suture was performed using polyglactin 910 4.0 (Vicryl®, Ethicon Endo-Surgery Inc., Greensboro, NC, USA). Immediately after the surgery, MTIs were loaded with a removable overdenture following the loading protocol under the Schnitman concept [16]. Then, patients were controlled at 2, 7, 30, 60, and 90 days.

After 8 months, the MTIs were extracted with a trephine bur of 3.5 mm in diameter, and the samples were conditioned for histomorphometric analysis by BS-SEM (Figure 1).

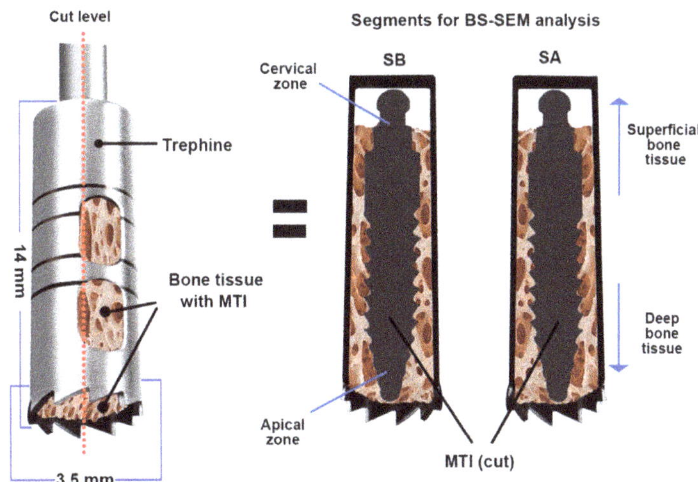

Figure 1. Bone sample with mini transitional implants (MTIs) obtained by trephine. A cut was made to obtain two samples (SA and SB) for backscattered scanning electron microscopy (BS-SEM) analysis. MTI = mini transitional implant; SA and SB = Segments A and B obtained after the cut.

2.2. Histomorphometric Analysis of Samples by BS-SEM

The MTIs were surgically removed using a trephine bur (implant plus its bone segment, Figure 1). They were subsequently fixed in 10% neutral buffered formaldehyde for 48 h. Later, the bone segments were prepared for their inclusion in light-curing resin (Technovit®). Posteriorly, the included samples were cut following the central longitudinal axis of the implant (Figure 1) using an Exact® diamond band saw (0.2 mm thick) and treated for BS-SEM analysis following the procedure described in the literature [17,18].

During BS-SEM analysis, several microphotographs were obtained (50× to 150×). The area of tissue between two threads of the implant surface was established as a study unit (Figure 2), referred to as the "bone area fraction occupancy" (BAFO). The following parameters were analyzed in the BAFOs:

1. Bone tissue classification present in the BAFO considering trabecular or cortical bone. The classification was determined considering the predominant form of bone present.
2. Bone to implant contact (%BIC): The contact between the bone-calcified tissue and implant surface is expressed as the BIC percentage according to the total implant surface between two threads (Figure 2).
3. Bone fill percentage (%Bf): This is the calcified tissue percentage in the BAFO interspersed by vascular and marrow spaces.
4. Lamellar bone (LB): This bone tissue type exhibits osteonal organization and is measured according to lamellar apposition (Figure 2).
5. Fibroreticular or woven bone (WB): This bone tissue type has a regular bone structure with isolated and polygonal cellular spaces (Figure 2).
6. Calcified chondroid tissue (ChT): This bone tissue type presents a characteristic aspect of calcified tissues with large, irregular, and confluent cellular spaces (Figure 2).
7. Vascular or medullar spaces: black spaces in the BAFO close to the calcified tissue compatible with spaces occupied by blood vessels or bone marrow (Figure 2).

In accordance with the variables described above, we performed quantitative valuation and qualitative analysis. Quantification of tissue areas was performed using the software ImageJ. The SPSS statistical software was used for the analysis of the quantitative results. The ANOVA or Kruskal–Wallis tests were applied and the p-value < 0.05 was considered significant.

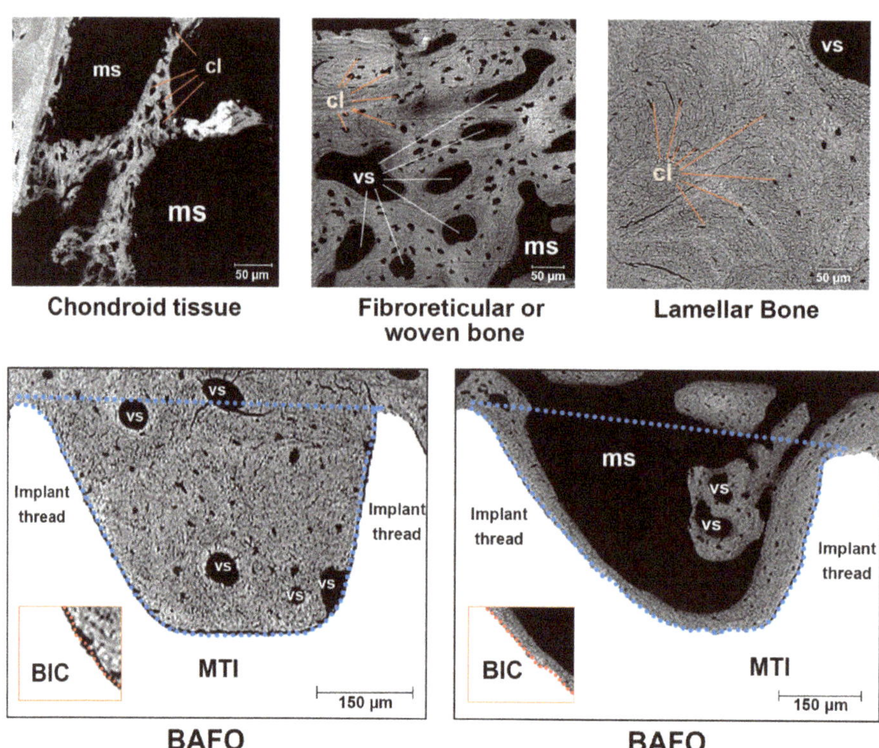

Figure 2. In the upper part, the types of bone tissue studied in the samples are shown; in the lower part, the bone area fraction occupancy (BAFO) used as an analysis unit in this study is shown (delimited by the blue broken line between two implant threads). Samples of Bone to Implant Contact (BIC) area are shown in the boxes inside the inferior images. cl = cellular lacunae; ms = medullar space; vs = vascular space; MTI = mini transitional implant (surface).

3. Results

3.1. General Aspects

No complications were observed or related by patients in the post-operative period. No mobility or lost implants were observed. A sample was lost during cutting and preparation for BS-SEM analysis. Consequently, 15 samples were studied.

3.2. Histomorphology Findings

The histology of the tissue found around the MTIs (BAFOs) obtained by BS-SEM is summarized in Tables 1 and 2 and Figures 3 and 4. In general, in the space between the MTI surface and trephine, we observed tissue with the characteristics of cortical bone close to the cervical area (near to the implant abutment, Figure 3(1,3). More trabecular bone was observed in the deep area (Figure 3(2,4)). A total of 306 BAFOs were observed, of which 243 BAFOs with calcified tissue were studied. The BAFOs without calcified tissue (63 BAFOs) were observed mainly in relation to the middle implant portion. The samples showed a predominance of trabecular bone. However, the highest BIC percentage considering all samples was associated with cortical bone. Table 1 shows the quantitative observations made in the samples considering the parameters of bone tissue characteristics, bone fill percentages, and BIC. Lamellar bone was the tissue most frequently observed, followed by woven bone. Regions with chondroid tissue were significantly less frequent (Table 2, Figure 4). Even when significant differences were observed in the amount of calcified tissue and vascular spaces between the samples, no significant differences were observed in the quantitative analysis considering lamellar, woven, and chondroid tissue (Figure 4). Consequently, the qualitative analysis revealed that the newly formed tissue in the BAFO had a regular cell distribution and characteristics of advanced maturation. In this region, several vascular spaces were observed with surrounding lamellar bone formation. In general, bone exhibited a tendency toward the mature osseous tissue. Besides, no remains of metal or particle corrosion detachment were observed by BS-SEM in the interface BIC or BAFO. Remodeled activity was observed, presumably depending on the forces applied to the implant.

Table 1. Summary of quantitative analysis of bone tissue observed in the BAFOs (Total samples = 15). For BAFO, %Bf, and %BIC definitions see the methodology segment. SD = standard deviation.

Bone Tissue	Number of BAFOs Analyzed (Total: 306)	%Bf in BAFO	%BIC (Total)
Mainly filled with cortical bone	94	37% to 100% (mean 62% SD 15.65)	63%
Mainly filled with trabecular bone	149	24% to 73% (mean 44% SD 14.01)	37%
Without calcified tissue	63	—	—

Table 2. Summary of quantitative analysis of samples considering bone tissue types. BAFOs with calcified tissues were considered (N = 243). SD = standard deviation.

Bone Tissue Types	Mean % (SD)
Lamellar bone (% of calcified tissues in BAFOs)	72.13% (13.58)
Woven bone (% of calcified tissues in BAFOs)	26.04% (12.39)
Chondroid bone (% of calcified tissues in BAFOs)	1.82% (3.69)
Medullar and vascular spaces (considering all BAFOs area)	39.41% (15.98)

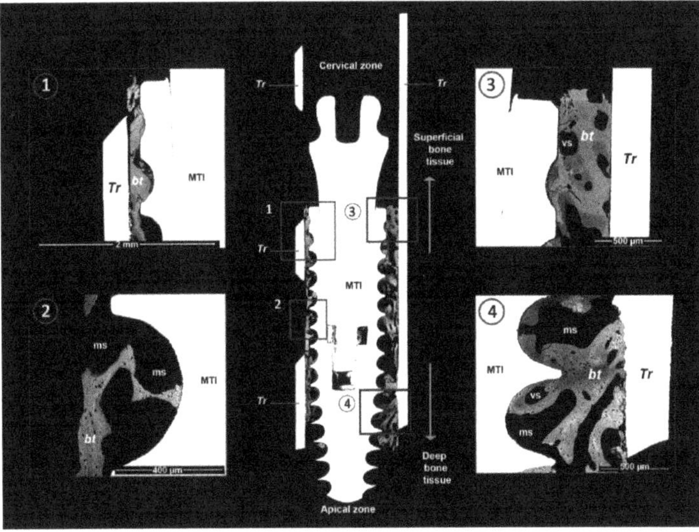

Figure 3. BS-SEM osseointegration between grade V titanium mini-transitional implants (MTI) and the bone surface. The BS-SEM image shows close contact between the implant surface and bone. Furthermore, bone tissue with an advanced degree of maturation and mainly lamellar and fibroreticular tissue organization is observed overall. The subfigures **1–4** show details of the BAFOs in the central image. MTI = mini transitional implant, Facility ®; Tr = trephine; bt = bone tissue; ms = medullar space; vs = vascular space.

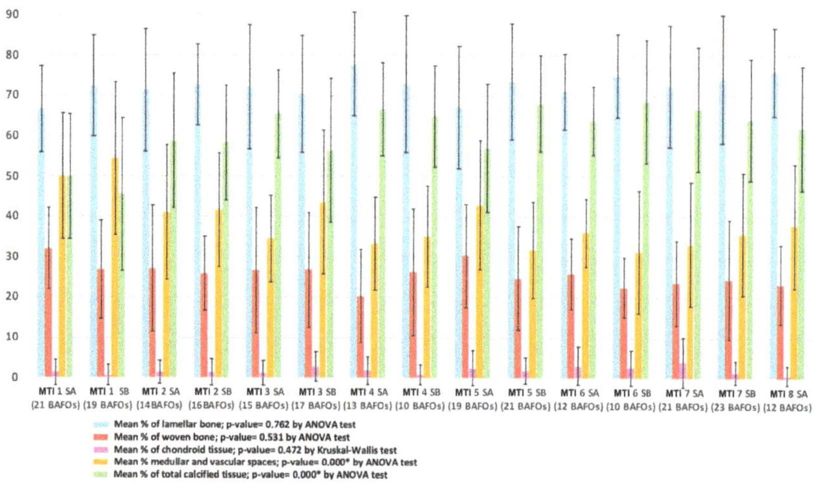

Figure 4. Distribution of calcified tissue considering each sample. MIT = mini transitional implant; SA and SB = Segment A and Segment B of sample (see Figure 1). The number of BAFOs analyzed is described below each set of bars belonging to a sample. No significant differences were observed in the distribution of the types of bone tissue. * Significant differences were observed in the distribution of vascular spaces and general calcified tissue between the samples.

4. Discussion

The results of the osseointegration process associated with MTIs placed in the anterior region of the mandible and immediately loaded by overdenture in humans show unprecedent evidence of the successful osseointegration of grade V titanium in the form of MTIs. The bone quality analyzed in the BAFOs showed characteristics of vascularized mature bone, with minimal evidence of bone in the resorption stage. This result is in agreement with the published histological findings in which the biocompatible and osteoconductive properties of substrate grade V titanium alloys were evidenced through in vivo studies [6,7,19].

The qualitative evaluation of histology indicated intimate contact between the bone and implant surface, especially in relation to cortical bone (Table 1). Bone formation around MTIs showed healing in the BAFO by an intramembranous-type bone healing pattern, and the appositional formation of new bone was observed where direct contact existed between the implant and bone immediately after placement. These findings are consistent with those described by Granato et al. [20], who observed BAFOs with woven bone appearing with a random orientation at 3 weeks, followed by the remodeling and replacement of woven bone by mature lamellar bone at 6 weeks.

The deposition of the bone matrix around the implant determines the success of the matrix, which depends directly on the process that induces the osteoblast proliferation, differentiation, secretion of extracellular matrix proteins, and tissue mineralization [21]. Along with the above, osseointegration is a dynamic process which alternates between bone formation and bone resorption. Consequently, the magnitude, direction, and period of any forces applied over the implant–bone zone will determine whether maintenance or failure of osseointegration equilibrium results [22]. The results presented here show that the properties of titanium V MTIs plus the correct surgical procedure for implant placement achieved adequate stability, allowing a successful osseointegration process, even when an immediate load was applied through an overdenture. It is logical that osseointegration may be influenced by the biomechanical properties, surface texture, and chemical composition of dental implants [23]. Thus, the implant surface (Neoporos) may be a factor that could improve implant success. Castro et al. [24] stated that the Neoporos process is a subtractive surface treatment consisting of zirconium oxide blasting and a series of acid etching baths. Currently, roughened surfaces have been used to increase the total surface area available for osteoconductive process, which could improve bone formation around the implants [25]. Consequently, subtractive treatment of surface and implant design have been considered critical variables that affect stability and quick osseointegration of immediately loaded implants. The above, without consideration of the additive treatment applying bioactive layers on the dental titanium surface, like those performed to improve bone regeneration around degradable temporary orthopedic implants [26]. However, the potential use of additive treatments on temporary implants destined to immediate load should be considered because the advances reported have shown promising results regarding the osseointegration in early stage, the quantity of bone to implant contact, and the adhesion strength of bone-to-implant [27]. Moreover, additive treatment by biodegradable bioactive nanofilms with a sub-micrometric thickness could improve the implant surface without affecting the primary stability required during its insertion. The nanofilms can promote cell proliferation and differentiation, and play a significant role in the matrix calcification [28].

The mechanical resistance of grade V titanium ensures the load transmission to the bone tissue over a prolonged period of time, which is important when lost hard tissues are replaced with a prosthesis [29]. Therefore, the use of alloys with improved mechanical properties in orthopedics and also in implantology is important, especially considering the increasing use and validation of short [30] and small-diameter implants [3] as well as the greater longevity of the patient population [6]. Compared with Co–Cr–Mo alloys, titanium alloy is almost twice as strong and has half the elastic modulus. Compared with 316 L stainless steel, grade V titanium alloy is roughly equal in strength; however, it has half the modulus. This titanium alloy is a material that is frequently used in the orthopedic area [21] mainly because of its superior mechanical properties.

Finally, the clinician must consider several aspects for implant selection, such as patient history (parafunctional habits, implant fracture, among others) or load conditions to which the implant will be subjected [31]. The use of titanium grade V in MTIs for immediate prosthetic loading has shown good biomechanical properties and successful osseointegration, maintaining a functional temporary overdenture during the study period.

5. Conclusions

Grade V titanium is an adequate material for dental implants from the point of view of its biocompatibility when considering the osseointegration process. Furthermore, the visualization of bone tissue after the removal of MTIs implants with trephine through histomorphometric analysis by BS-SEM demonstrated the presence of mature bone tissue that was mostly formed by lamellar bone. This adequate quality of bone tissue surrounding the dental implant and the intimate and stable contact between the implant surface and bone tissue revealed that, despite the immediate occlusal load applied to the implant, titanium V MTI was adequately integrated into the bone tissue, ensuring the stability of the temporary prosthesis.

Author Contributions: Conceptualization: V.B. and R.L. Research, methodology and supervision: V.B., R.L., B.W. and I.V.-G.; Validation: V.B., R.L., and I.V.-G.; Original draft: V.B., I.V.-G., R.L., C.S., N.F., P.A.-M., and M.-C.M.; Image and photo work: I.V.-G., P.A.-M.; Composition: V.B., I.V.-G., R.L., C.S., N.F., P.A.-M., B.W. and M.-C.M.; Writing, review and editing: V.B., I.V.-G., R.L., C.S., N.F., P.A.-M., B.W. and M.-C.M. All authors have read and agreed to the published version of the manuscript.

Funding: This research was funded in part by MINEDUC-UA project, code ANT 1855, and Insertion Program DI16-6009, Universidad de La Frontera.

Acknowledgments: We acknowledge ILAPEO (Curitiba, Brazil) for their support of the implants. We also appreciate the collaboration of the Implant Clinic of the School of Dentistry, Universidad Mayor, Santiago, Chile.

Conflicts of Interest: The authors declare no conflict of interest.

References

1. Mundt, T.; Schwahn, C.; Stark, T.; Biffar, R. Clinical response of edentulous people treated with mini dental implants in nine dental practices. *Gerodontology* **2015**, *32*, 179–187. [CrossRef] [PubMed]
2. Veltri, M.; Ferrari, M.; Balleri, P. One-year outcome of narrow diameter blasted implants for rehabilitation of maxillas with knife-edge resorption. *Clin. Oral Implants Res.* **2008**, *19*, 1069–1073. [CrossRef] [PubMed]
3. Assaf, A.; Saad, M.; Daas, M.; Abdallah, J.; Abdallah, R. Use of narrow-diameter implants in the posterior jaw: A systematic review. *Implant Dent.* **2015**, *24*, 294–306. [CrossRef]
4. De Almeida, F.D.; Carvalho, A.C.; Fontes, M.; Pedrosa, A.; Costa, R.; Noleto, J.W.; Mourao, C.F. Radiographic evaluation of marginal bone level around internal-hex implants with switched platform: A clinical case report series. *Int. J. Oral Maxillofac. Implants* **2011** *26*, 587–592. [PubMed]
5. Calvo-Guirado, J.L.; Gómez-Moreno, G.; Aguilar-Salvatierra, A.; Guardia, J.; Delgado-Ruiz, R.A.; Romanos, G.E. Marginal bone loss evaluation around immediate non-occlusal microthreaded implants placed in fresh extraction sockets in the maxilla: A 3-year study. *Clin. Oral Implants Res.* **2015**, *26*, 761–767. [CrossRef]
6. Diaz-Sanchez, R.-M.; de-Paz-Carrion, A.; Serrera-Figallo, M.-A.; Torres-Lagares, D.; Barranco, A.; León-Ramos, J.-R. Gutierrez-Perez, J.-L. In Vitro and In Vivo Study of Titanium Grade IV and Titanium Grade V Implants with Different Surface Treatments. *Metals* **2020**, *10*, 449. [CrossRef]
7. Imam, M.; Fraker, A. Titanium Alloys as Implant Materials. In *Medical Applications of Titanium and Its Alloys: The Material and Biological Issues*; Lemons, J., Brown, S., Eds.; ASTM International: West Conshohocken, PA, USA, 1996; pp. 3–16.
8. Sharan, J.; Lale, S.V.; Koul, V.; Mishra, M.; Kharbanda, O.P. An Overview of Surface Modifications of Titanium and its Alloys for Biomedical Applications. *Trends Biomater. Artif. Organs* **2015**, *29*, 176–187.
9. Ribeiro da Silva, J.; Castellano, A.; Malta Barbosa, J.P.; Gil, L.F.; Marin, C.; Granato, R.; Bonfante, E.A.; Tovar, N.; Janal, M.N. Coelho, P.G. Histomorphological and histomorphometric analyses of grade IV commercially pure titanium and grade V Ti-6Al-4V titanium alloy implant substrates: An in vivo study in dogs. *Implant Dent.* **2016**, *25*, 650–655. [CrossRef]
10. Johansson, C.B.; Han, C.H.; Wenneberg, A.; Albrektsson, T. A quantitative comparison of machined commercially pure titanium and titanium-aluminum-vanadium implants in rabbit bone. *Int. J. Oral Maxillofac. Implants* **1998**, *13*, 315–321.
11. Han, C.H.; Johansson, C.B.; Wenneberg, A.; Albrektsson, T. Quantitative and qualitative investigations of surface enlarged titanium and titanium alloy implants. *Clin. Oral Implants Res.* **1998**, *9*, 1–10. [CrossRef]
12. Saulacic, N.; Bosshardt, D.D.; Bornstein, M.M.; Berner, S.; Buser, D. Bone apposition to a titanium-zirconium alloy implant as compared to two other titanium-containing implants. *Eur. Cells Mater.* **2012**, *23*, 273–286. [CrossRef] [PubMed]
13. Khan, M.A.; Williams, R.L.; Williams, D.F. Conjoint corrosion and wear in titanium alloys. *Biomaterials* **1999**, *20*, 765–772 [CrossRef]
14. Pennekamp, P.H.; Gessmann, J.; Diedrich, O.; Burian, B.; Wimmer, M.A.; Frauchiger, V.M.; Kraft, C.N. Short-term microvascular response of striated muscle to cp-Ti, Ti-6Al-4V, and Ti-6Al-7Nb. *J. Orthopaedic Res. Off. Publ. Orthopaedic Res. Soc.* **2006**, *24*, 531–540 [CrossRef] [PubMed]
15. Manzanares, M.C.; Franch, J.; Carvalho, P.; Belmonte, A.M.; Tusell, J.; Franch, B.; Fernandez, J.M.; Cleries, L.; Morenza, J.L. BS-SEM evaluation of the tissular interactions between cortical bone and calcium-phosphate covered titanium implants *Bull. Group Int. Rech. Sci. Stomatol. Odontol.* **2001**, *43*, 100–108. [PubMed]
16. Schnitman, P.A.; Wohrle, P.S.; Rubenstein, J.E. Immediate fixed interim prostheses supported by two-stage threaded implants: Methodology and results. *J. Oral Implantol.* **1990**, *16*, 96–105. [PubMed]
17. Manzanares, M.C.; Calero, M.I.; Franch, J.; Jiménez, M.P.; Serra, I. Optimisation of a scheduled study for undecalcified samples. *Microsc. Anal.* **1997**, *50*, 17–19.
18. Franch, J.; Pastor, J.; Franch, B.; Durall, I.; Manzanares, M.C. Back-scattered electron imaging of a non-vertebral case of hypervitaminosis A in a cat. *J. Feline Med. Surg.* **2000**, *2*, 49–56. [CrossRef]
19. Castellano, A.; Gil, L.F.; Bonfante, E.A.; Tovar, N.; Neiva, R.; Janal, M.N.; Coelho, P.G. Early healing evaluation of commercially pure titanium and Ti-6Al-4V presenting similar surface texture: An in vivo study. *Implant Dent.* **2017**, *26*, 338–344. [CrossRef]
20. Granato, R.; Bonfante, E.; Castellano, A.; Khan, R.; Jimbo, R.; Marin, C.; Morsi, S.; Witek, L.; Coelho, P.G. Osteointegrative and microgeometric comparison between micro-blasted and alumina blasting/acid etching on grade III and V titanium alloys (Ti-6Al-4V). *J. Mech. Behav. Biomed. Mater.* **2019**, *97*, 288–295. [CrossRef]
21. Zhang, H.; Lewis, C.G.; Aronow, M.S.; Gronowicz, G.A. The effects of patient age on human osteoblasts' response to Ti-6Al-4V implants in vitro. *J. Orthop. Res.* **2004**, *22*, 30–38. [CrossRef]
22. Delgado-Ruiz, R.A.; Calvo-Guirado, J.L.; Romanos, G.E. Effects of occlusal forces on the peri-implant-bone interface stability. *Periodontology 2000* **2019**, *81*, 179–193. [CrossRef] [PubMed]
23. Colnot, C.; Romero, D.M.; Huang, S.; Rahman, J.; Currey, J.A.; Nanci, A.; Brunski, J.B.; Helms, J.A. Molecular analysis of healing at a bone-implant interface. *J. Dent. Res.* **2007**, *86*, 862–867. [CrossRef] [PubMed]

24. Castro, D.S.; Araujo, M.A.; Benfatti, C.A.; Araujo Cdos, R.; Piattelli, A.; Perrotti, V.; Lezzi, G. Comparative histological and histomorphometrical evaluation of marginal bone resorption around external hexagon and Morse cone implants: An experimental study in dogs. *Implant Dent.* **2014**, *23*, 270–276. [CrossRef] [PubMed]
25. Shi, G.S.; Ren, L.F.; Wang, L.Z.; Lin, H.S.; Wang, S.B.; Tong, Y.Q. H_2O_2/HCl and heat-treated Ti-6Al-4V stimulates pre-osteoblast proliferation and differentiation. *Oral Surg. Oral Med. Oral Pathol. Oral Radiol. Endod.* **2009**, *108*, 368–375. [CrossRef] [PubMed]
26. Li, B.; Gao, P.; Zhang, H.; Guo, Z.; Zheng, Y.; Han, Y. Osteoimmunomodulation, osseointegration, and in vivo mechanical integrity of pure Mg coated with HA nanorod/pore-sealed MgO bilayer. *Biomater. Sci.* **2018**, *6*, 3202–3218. [CrossRef]
27. Shi, R.; Hayashi, K.; Ishikawa, K. Rapid Osseointegration Bestowed by Carbonate Apatite Coating of Rough Titanium. *Adv. Mater. Interfaces* **2020**, *7*, 2000636. [CrossRef]
28. Vannozzi, L.; Gouveia, P.; Pingue, P.; Canale, C.; Ricotti, L. Novel Ultrathin Films Based on a Blend of PEG-b-PCL and PLLA and Doped with ZnO Nanoparticles. *ACS Appl. Mater. Interfaces* **2020**, *12*, 21398–21410. [CrossRef]
29. Bauer, S.; Schmuki, P.; von der Mark, K.; Park, J. Engineering biocompatible im- plant surfaces Part I: Materials and surfaces. *Prog. Mater. Sci.* **2013**, *58*, 261–326. [CrossRef]
30. Kotsovilis, S.; Fourmousis, I.; Karoussis, I.K.; Bamia, C. A systematic review and meta-analysis on the effect of implant length on the survival of rough-surface dental implants. *J. Periodontol.* **2009**, *80*, 1700–1718. [CrossRef]
31. McCracken, M. Dental Implant Materials: Commercially Pure Titanium and Titanium Alloys. *J. Prosthodontics* **1999**, *8*, 40–43. [CrossRef]

Article

Effects of Ti6Al4V Surfaces Manufactured through Precision Centrifugal Casting and Modified by Calcium and Phosphorus Ion Implantation on Human Osteoblasts

Fiedler Jörg [1], Katmer Amet Betül [1], Michels Heiner [2], Kappelt Gerhard [3] and Brenner Rolf Erwin [1,*]

[1] Department of Orthopedic Surgery, Division for Biochemistry of Joint and Connective Tissue Diseases, Ulm University, 89081 Ulm, Germany; joerg.fiedler@uni-ulm.de (F.J.); betuel.katmer@uni-ulm.de (K.A.B.)
[2] Access e.V., 52072 Aachen, Germany; h.michels@access-technology.de
[3] Peter Brehm GmbH, 91085 Weisendorf, Germany; gerhard.kappelt@peter-brehm.de
* Correspondence: rolf.brenner@uni-ulm.de; Tel.: +49-731-500-63281

Received: 30 October 2020; Accepted: 9 December 2020; Published: 16 December 2020

Abstract: (1) In order to enable a more widespread use of uncemented titanium-based endoprostheses to replace cobalt-containing cemented endoprostheses for joint replacement, it is essential to achieve optimal osseointegrative properties and develop economic fabrication processes while retaining the highest biomedical quality of titanium materials. One approach is the usage of an optimized form of Ti6Al4V-precision casting for manufacturing. Besides the chemical and physical properties, it is necessary to investigate possible biological influences in order to test whether the new manufacturing process is equivalent to conventional methods. (2) Methods: Primary human osteoblasts were seeded on discs, which were produced by a novel Ti6Al4V centrifugal-casting process in comparison with standard machined discs of the same titanium alloy. In a second step, the surfaces were modified by calcium or phosphorus ion beam implantation. In vitro, we analyzed the effects on proliferation, differentiation, and apoptotic processes. (3) Results: SEM analysis of cells seeded on the surfaces showed no obvious differences between the reference material and the cast material with or without ion implantation. The MTT (3-(4,5-dimethylthiazol-2-yl)-2,5-diphenyltetrazolium bromide) proliferation assay also did not reveal any significant differences. Additionally, the osteogenic differentiation process tested by quantitative polymerase chain reactions (PCR), Alizarin red S assay, and C-terminal collagen type I propeptide (CICP) Elisa was not significantly modified. No signs of induced apoptosis were observed. (4) Conclusions: In this study, we could show that the newly developed process of centrifugal casting generated a material with comparable surface features to standard machined Ti6Al4V material. In terms of biological impact on primary human osteoblasts, no significant differences were recognized. Additional Ca- or P-ion implantation did not improve or impair these characteristics in the dosages applied. These findings indicate that spin casting of Ti6Al4V may represent an interesting alternative to the production of geometrically complex orthopedic implants.

Keywords: Ti6Al4V; centrifugal casting; ion implantation; human osteoblast

---

1. Introduction

Nowadays, there is high demand for cost-effective knee joint endoprosthesis in orthopedic arthroplasty. The currently used respective Gold standard material for cemented endoprosthesis, Co28Cr6Mo, however, is in discussion in terms of biocompatibility and longevity because of its cobalt content. Ti6Al4V is the most frequently used titanium-alloy for uncemented orthopedic endoprosthesis implantation, but fabrication of standard 3-dimensional structures for joint arthroplasty involves cost intensive usage of machining blanks and milling machines to remove projecting titanium parts [1].

Therefore, a cost-effective titanium casting process could achieve improvement in several ways. It could lead to the partial replacement of cobalt-containing implants and avoid the use of bone cement in these cases based on superior features of osseointegration. However, because of the high melting point of titanium and its unfavorable fluidity and reactivity in a molten state, this fabrication procedure is not routinely used so far [2,3].

As described previously, an optimized centrifugal casting process, including casting and cooling conditions, crucible and mold material, and enhanced heat treatment facilitates, was used for the manufacturing of near net-shape titanium-based implants (Figure 1 is an illustration of a centrifugal precision casting device from Michels, Aachen, Germany) [4]. In the applied casting process on the Leicomelt 5 TP casting device, the solid metal alloy is placed in a cold wall crucible. This type of crucible consists of a ring-shaped wall of water-cooled palisades made from copper. A magnetic alternating field is induced by the application of current to an induction coil surrounding the palisade package. This field generates heat in the alloy material due to ohmic losses, eventually melting the solid alloy. Melting and subsequent casting is carried out under an inert gas/vacuum atmosphere. The liquid metal is poured by tilting the crucible through a separation valve between the melting and casting chamber into a heating box containing the casting setup consisting of sprue, melt distributor and ceramic shell mold. The heat box and casting setup are fixed on a rotating table inside the casting chamber. Prior to the pouring of the melt, the casting the table is set into a rotation of up to 400 RPM, creating rotational forces on the liquid metal as it enters the casting setup. Following the applied force, the melt flows into the ceramic shell mold, thus filling it within 1.5 s. The cooled down metal is manually freed from the ceramic shell mold, and the in vitro samples are cut by a water jet.

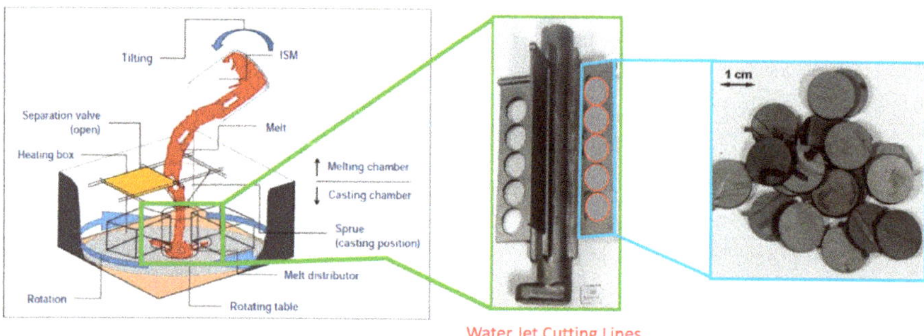

Figure 1. Principal centrifugal casting layout (Leicomelt TP5) (**left**); cast part wax model providing 4 rectangular plates, water jet cutting lines in red (**center**); water-jet cut of in vitro test specimen (**right**). Discs were 14 mm in diameter and 2 mm in height.

Besides the complex manufacturing process of titanium-based implants, it is necessary to focus on their biological aspects. Stability after uncemented implantation into bone depends on a proper interaction between the material and the cells of the surrounding tissue, especially osteoblasts [5]. While pure titanium or titanium-based alloys like Ti6AL4V are usually regarded as having excellent mechanical and biocompatibility properties, several organic and inorganic surface modifications were used to further enhance osseointegration [6]. With respect to inorganic components, besides surface modification by the addition of hydroxyapatite, complex 3-dimensional structures were generated by plasma treatment in order to deposit calcium or phosphorus ions onto the surfaces [7,8]. Since delamination of hydroxyapatite coatings with negative long-term effects has been reported [8,9], ion implantation of calcium and phosphorus into the material surface has been tested, encouraging the formation of calcium phosphate precipitates [10,11]

The alloy surface should maintain the cell adhesion, proliferation and differentiation processes of osteoblasts, which are dependent on biochemical, topographical and biomechanical parameters. One example is the increasing response of different cells to materials using nano topography by supporting respective adhesion and proliferation [12,13]. With respect to osseointegration, the material should not decrease the expression of necessary osteogenic differentiation factors like RUNX2 (Runt-related transcription factor 2) or the synthesis of collagen type I, which is essential for extracellular matrix synthesis [14].

The aim of this in vitro study was to investigate the possibilities of using machine blanks and milling machines to transfer a medical grade process of knee endoprosthesis production into a more cost-effective, less material consuming optimized centrifugal casting process, without worsening the biological effects on primary human osteoblasts. We used a modified manufacturing technique for the production of near net-shape precision centrifugal castings of Ti6Al4V, representing an alloy widely used for medical treatment. In addition, a calcium and phosphorus ion beam implantation into the implant's surface was tested to see if it modified the biological outcome.

2. Materials and Methods

An overview of the experimental setup is given in Figure 2. Besides the used test specimen, the cell biological tests are illustrated.

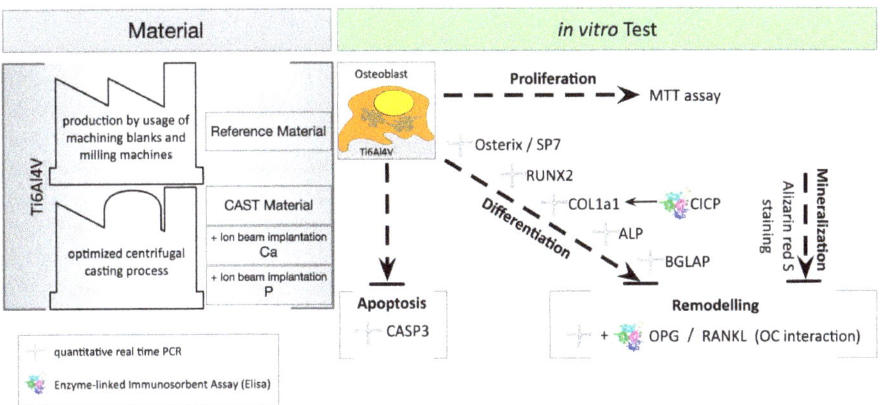

Figure 2. Scheme of the materials and methods used for in vitro testing. The medical grade alloy Ti6Al4V was processed under standard conditions (reference material; REF) or in the optimized centrifugal casting process (CAST). Some specimens were modified by ion beam insertion of Ca- or P-ions into the surface. Discs of 14 mm in diameter were seeded with human primary osteoblasts and underwent different in vitro testing. Cell biologic analysis focused on cell proliferation, osteogenic differentiation, matrix mineralization, remodeling processes, and apoptosis.

MTT: colorimetric assay of cellular metabolic activity by reducing the tetrazolium dye MTT 3-(4,5-dimethylthiazol-2-yl)-2,5-diphenyltetrazolium bromide to its insoluble formazan. Alizarin red S: staining of the calcified matrix that is synthesized by the osteoblasts. Quantitative real time PCR: gene expression analysis of differentiation markers. Osterix/SP7: a transcription factor that is highly conserved among bone-forming cells and responsible for osteogenic differentiation. RUNX2: Runt-related transcription factor 2 associated with osteoblast differentiation. COL1a1: collagen type I alpha 1 as one chain of collagen type I—the major structural protein of bone. CICP: type I C-terminal collagen propeptide. ALP: tissue non-specific Alkaline phosphatase. BGLAP: osteocalcin, also known as bone gamma-carboxyglutamic acid-containing protein, is a calcium binding protein. CASP3: caspase-3 plays an important role in cell apoptosis. OPG: osteoprotegerin, a protein produced by osteoblasts to counteract bone resorption by osteoclasts. RANKL: the receptor activator of nuclear factor kappa-B ligand, stimulator of bone resorption. OC: osteoclasts.

2.1. Primary Human Osteoblasts

Primary human osteoblasts were isolated from human bone specimens collected from routine joint replacement surgery with the consent of patients and after approval from the local ethics committee of the University of Ulm. Overall, samples from 7 donors were used. Cells were isolated and cultivated under standard conditions as described previously [15].

2.2. Test Specimen

For the in vitro testing, discs of the respective materials with a diameter of 14 mm and a height of 2 mm were used. The reference test specimens (REF) consisted of machined, aluminum-oxide-blasted Ti6Al4V and were manufactured by Peter Brehm GmbH (Weisendorf, Germany). The spin-cast discs were manufactured by the optimized centrifugal precision-casting technique, which was developed in cooperation between the independent research facility Access e. V. (Aachen, Germany) and the implant manufacturer, Peter Brehm GmbH (Weisendorf, Germany), as described earlier [6]. Following the casting process, the parts were heat HIPed (hot-isostatic pressed) according to Ti6Al4V standard procedures and parameters (1000 bar at 920 °C for 120 min). During the HIP procedure, by parallel application of temperature and pressure, the porosity in the cast part is effectively reduced, increasing the density of the cast material by significantly decreasing the volume fraction of possibly existing casting defects. The test specimens created by the melting cast procedure (CAST) were cut out of bigger sized plates to maintain the typical surface features generated by the casting process. Finally, the CAST discs were aluminum-oxide-blasted, using the same standard process as for the reference material [6].

Ion beam implantation was used to modify the properties of material surfaces [15]. Ca- or P-atoms were ionized and accelerated in electric fields and thus implanted into the surface of planar Ti6Al4V seals. The implantation was carried out with a conventional low-energy implant DANFYSIK 1050 (Danfsik, Taastrup, Denmark) at the Helmholtz-Zentrum Dresden-Rosssendorf (HZDR), as described previously for Cu- and Ag-ions [15]. The surfaces were implanted with an energy of 30 keV and a dose of 1×10^{-16} cm^2 Ca- or P-ions.

2.3. SEM Analysis of Cell Adhesion

The surface structure of the discs and cell adhesion were investigated by using scanning electron microscopy Hitachi S-5200 (Hitachi, Tokyo, Japan) at the electron microscopy core facility of the University of Ulm, Germany. Specimens without cells were sputtered with gold-palladium (20 nm) under standard conditions. Specimens with cells were first fixed with 2.5% glutaraldehyde and 1% saccharose in 0.1 M of phosphate buffer before sputtering with gold-palladium.

2.4. Molecular Biological Methods

Gene expression analysis was performed with standard methods, as described earlier [16]. Human osteoblasts were seeded at a density of 20,000 cells per disc, which was placed in one well of a 24-well plate and covered with 1 ml of Dulbecco's Modified Eagle Medium (DMEM), 10% fetal calf serum (FCS), 100 µL penicillin/streptomycin solution, and 2 mM L-glutamine (all Biochrome, Berlin, Germany). For differentiation processes, 0.1 µM dexamethasone, 10 mM β-glycerophosphate, and 0.2 mM L-Ascorbic acid (all Sigma-Aldrich, Steinheim, Germany) were added to the medium.

For normalization of the quantitative real-time polymerase chain reaction (PCR) results, a cell sample of osteoblasts before seeding on the titanium surfaces was used in all experiments. The cell culture supernatant was used for Enzyme Linked Immuno Sorbant Assay (Elisa). The cells were lysed by using the RNeasy kit from Qiagen (Qiagen, Hilden, Germany), following company instructions. Synthesis of cDNA was carried out with the Omniscript kit from Qiagen (Qiagen, Hilden, Germany) in accordance with the given manuals. Gene expression was analyzed using a TaqMan StepOne Plus

(Life Technologies, Darmstadt, Germany) and ready-to-use TaqMan probes, which are commercially available from Life Technologies (listed in Table 1). Amplifications were carried out using the TaqMan Fast Advanced Master Mix (Life Technologies, Darmstadt, Germany). Gene expressions were calculated with the ΔΔCt method and normalized to HPRT1 as a housekeeping gene [17].

Table 1. The TaqMan probes and targets used. Probes were designed and tested for specificity by Life Technologies (Darmstadt, Germany).

Gene	Gene Accession No.	TaqMan Assay ID	Common Name
ALP	NM_000478.5	Hs01029144_m1	Alkaline phosphatase
BGLAP	NM_199173.5	Hs01587814_g1	Bone gamma-carboxyglutamate protein
CASP3	NM_004346.3	Hs00234387_m1	Caspase 3
COL1a1	NM_000088.3	Hs00164004_m1	Collagen type I alpha 1
HPRT1	NM_000194.2	Hs02800695_m1	Hypoxanthin-Phosphoribosyl-Transferase 1
OPG	NM_002546.3	Hs00900358_m1	Osteoprotegerin
RUNX2	NM_001015051.3	Hs00231692_m1	Runt-related transcription factor 2
TNF	NM_000594.3	Hs01113624_g1	Tumor necrosis factor
RANKL	NM_003701.3	Hs00243522_m1	Receptor activator of nuclear factor kappa-B ligand
SP7	NM_001173467.2	HS001866874_s1	Transcription factor Sp7, Osterix

Gene and Gene Accession No. from GeneBank, NIH, TaqMan Assay ID given identification No from Life Technologies.

2.5. Elisa

In order to analyze protein secretion, a human Osteoprotegerin Instant ELISA (affymetrix, San Diego, CA, USA) and a MicroVue® Bone CICP EIA Elisa (Quidel, San Diego, CA, USA) were used according to the manufacturers' instructions. The Elisas were analyzed using a Tecan Infinite M200 Pro (Tecan, Männedorf, Switzerland) for readout and the software iControl™ V. 1.2 or Magellan™ V. 7.2 for, all from Tecan, Männedorf, Switzerland).

2.6. MTT Assay

Cell proliferation was tested according to DIN ISO 10993-5 by using tetrazolium salt MTT (3-(4,5Dimethylthiazol-2-yl)-2,5-diphenyltetrazoliumbromid) as described elsewhere. Staining was conducted 24 h, 3 days, and 7 days after cell seeding on the specimen, which was cultivated under standard conditions [15]. At the indicated time points, the discs were placed into fresh wells of a 24-well plate in order to only include cells seeded on the titanium surface. The wells were filled with 1 mL MTT solution and incubated for 3 h. Then, the MTT solution was replaced by 200 μL of a 0.04 M HCl/Isopropanol mixture. The discs were reversed using forceps and sonicated for 3 min. The staining was analyzed with a Tecan Infinite M200 Pro (570 nm/650 nm), as described above (Section 2.5).

2.7. Alizarin Red S Staining

To prove the matrix calcification of the seeded specimens, Alizarin red S staining was performed. Therefore, 20,000 cells per disc were cultivated for 14 days in basal medium with the additional supplements as indicated above (Section 2.4). Then, the discs were placed into a new well and washed three times with 1 mL phosphate buffered saline (Biochrome, Berlin, Germany). Subsequently, 1 mL of 70% ethanol was added and incubated for 1 h. Thereafter, 1 mL of a 40 nM Alizarin staining solution was added and incubated on a horizontal shaker for 10 min. After several washing steps with pure water, an incubation for 15 min with 10% cetylpyrimidinchloride was performed. 100 μL was used to analyze the concentration of Alizarin red S using a Tecan Infinite M200 Pro plate reader (absorbance 652 nm), as described above (Section 2.5).

2.8. Statistical Testing

Statistical analysis was carried out using GraphPad Prism version 8.4 for Macintosh (GraphPad Software, La Jolla, San Diego, CA, USA, (www.graphpad.com)). Depending on the experimental setup, a one-way-Analysis of Variance (ANOVA) or two-way-ANNOVA was used. A Tukey post hoc test was performed for one-factorial analysis of variance; a Bonferroni post hoc test was used for two-factorial analysis. The level of significance was set to $p < 0.05$. The data are presented by a Scatter-dot plot with Median and interquartile range.

3. Results

Before cutting the test specimen, each plate was checked for the existence of Alpha Case, a critical surface feature, which can develop during the cooling phase of the casting process due to the presence of oxygen. While thin layers can be removed by etching, generally avoiding the formation of Alpha Case during the casting process is the preferred solution with respect to cost-effective part production. Figure 3 shows typical analyses of cast part samples before (Figure 3A,C) and after (Figure 3B,D) optimization of the employed centrifugal casting process, indicating that Alpha Case formation could be completely prevented in the test samples used for further analyses (Figure 3B,D).

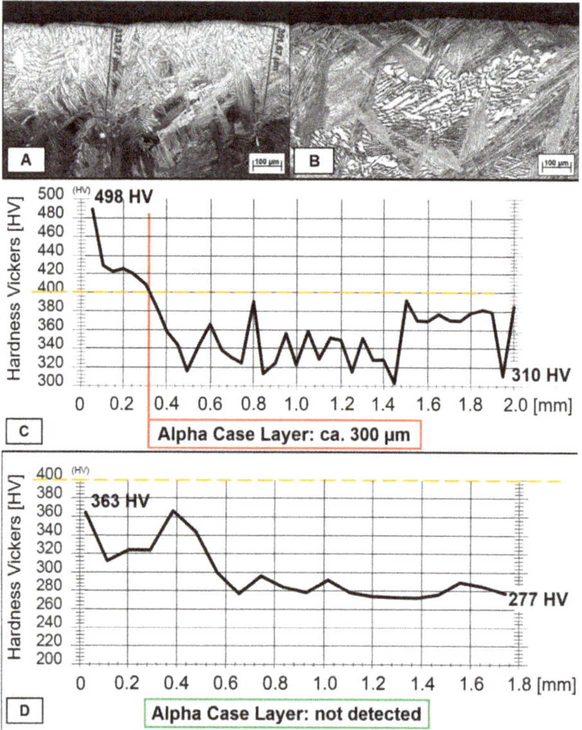

Figure 3. Microstructure analysis (Light Microscopy 100× magnification; Micro-Hardness) (**A**,**C**) Microstructure before process optimization, thickness of Alpha Case Layer critical (>15 µm); (**B**) and (**D**) Microstructure after process optimization, no Alpha Case detectable (Hardness Vickers, Load: HV0.10, Obj. 50×).

The chemical composition of the test specimen was checked by Inductively Coupled Plasma-Optical Emission Spectrometry (ICP-OES) analysis according to DIN 51008-2/DIN 510009. ICP-OES is an elemental analysis method to determine sample compositions on trace-level. Samples are water dissolved and conducted through a nebulizer into a spray chamber. The resulting aerosol is lead through an argonized plasma chamber operating at around 6000 to 7000 K. The aerosol takes up thermal energy and atomization, and ionization takes place. Electrons reach a higher state for a short amount of time and eventually drop back to ground level energy. While dropping back, energy is liberated as light waves (photons). For each element, the wavelength of the emitted light is characteristic and used to identify the elements present in the sample. Information regarding the content of an element is provided by the measured intensity of the detected wave lengths. Based on the machine calibration, these values are used to calculate the concentration of elements contained in the sample.

Representative results are shown in Table 2. The element amounts were within the specified boundaries and the results show a typical decrease of Al and an increase of elements such as Cu, Fe and Ti, respectively. This effect commonly occurs when a melting alloy containing Al is used at high temperatures. However, this effect is minimal for the Leicomelt 5 TP device, as shown by the results.

Table 2. Element concentration as specified for Ti6Al4V ELI and measured for the alloy material before and after the melting+casting+HIP process chain. Tests were conducted by Elektrowerk Weisweiler GmbH (Eschweiler-Weisweiler, Germany). Results are shown in weight percent (wt.%).

Element	Ti6Al4V ELI Specification	wt.% before Processing	wt.% after Melting, Casting & HIP
C	0.1 (max)	0.011	0.053
V	3.5–4.5	4.53	4.43
Al	5.5–6.75	6.18	6.03
O	0.2 (max)	0.176	0.131
N	0.05 (max)	0.005	0.001
Fe	0.3 (max)	0.215	0.223
H	0.015 (max)	0.005	0.012
Y	-	<0.001	<0.002
Ti	Balance	Balance	Balance

3.1. *Surface Characteristics and SEM Analysis*

The biocompatibility of medical implants is determined by several factors. Although the reference and cast specimens used in this study were made of the same titanium alloy and had a comparable major chemical composition as mentioned above, effects on the cell adhesion, proliferation, and differentiation of human osteoblasts could be influenced by the respective surface profile. Therefore, surface micro-topography was characterized.

When analyzing the average roughness Ra and Rz by using standard procedures, no significant differences were detected. For the reference surfaces, the mean value of Ra was 5.1 μm (SD 0.44 μm), while a Rz of 29.78 μm (SD 1.71 μm) was measured. For the cast samples, the mean value determined for Ra was 4.55 μm (SD 0.14 μm) and Rz ranged from 28.05 μm (SD 2.18 μm). In ion implanted surfaces, samples with implanted Ca-ions showed mean values of Ra 4.4.5 μm (SD 0.43 μm) and Rz 27.20 μm (2.53 μm). In cast specimens with implanted P-ions, Ra was measured with 4.86 μm (SD 0.45 μm), and Rz 29.58 μm (SD 2.07 μm) was detectable.

By using scanning electron microscopy, we could further show that the surfaces of both types of specimen were similar and that the seeded human osteoblasts could adhere and cover the surface after several days, as shown in Figure 4. Additionally, no different seeding behavior was observed after the ion beam implantation of calcium or phosphorus ions into the surface. The osteoblasts were comparable in terms of their cellular morphology and size.

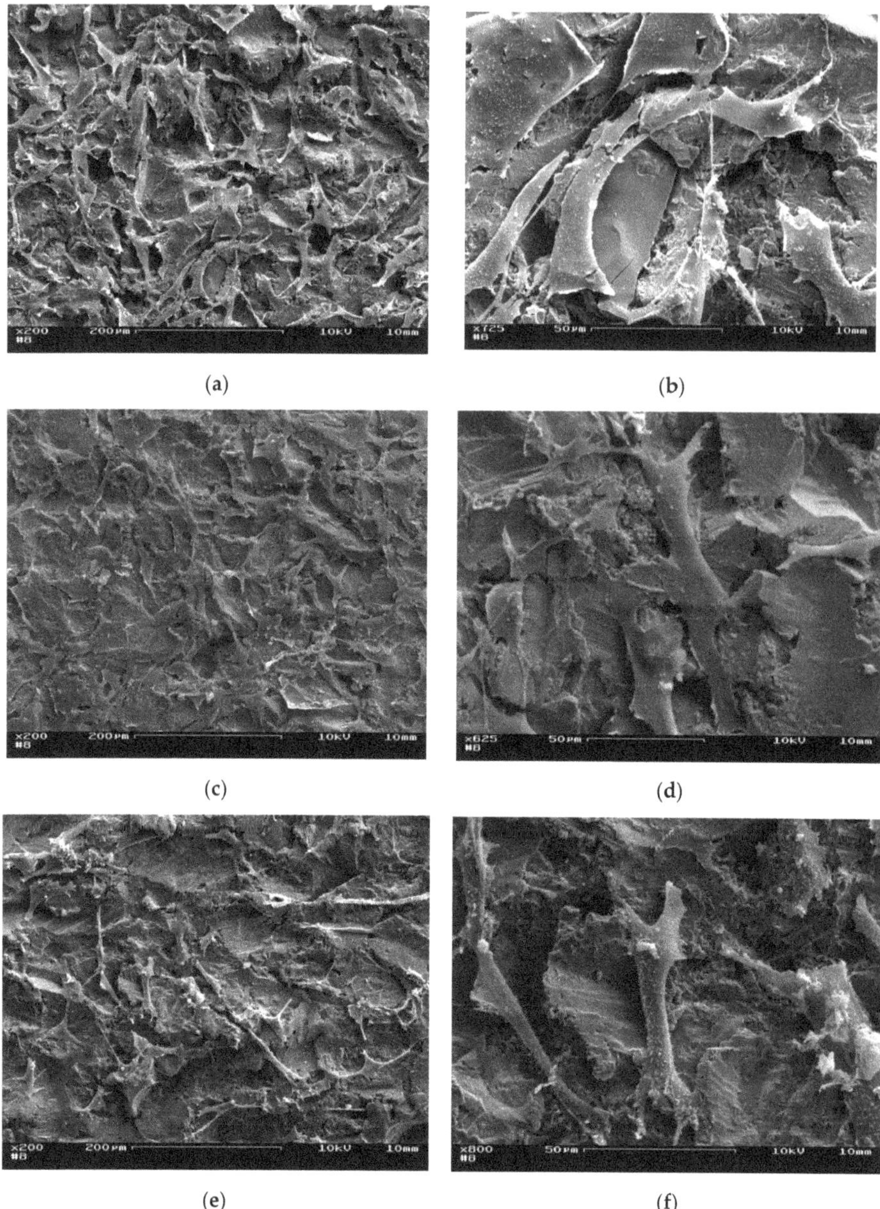

Figure 4. *Cont.*

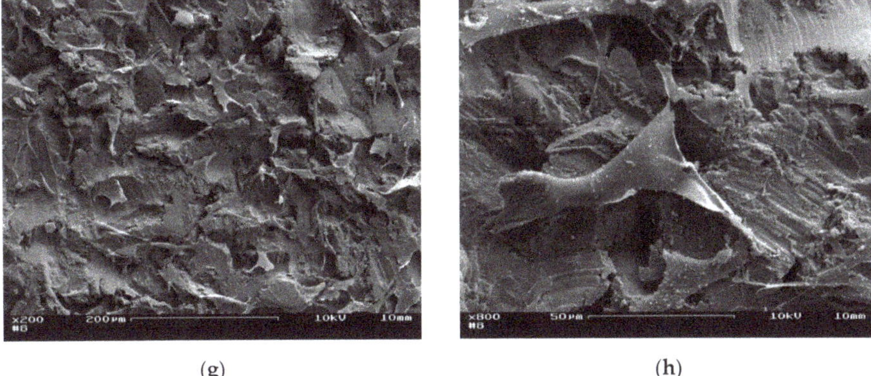

(g) (h)

Figure 4. SEM images of cell seeded surfaces. On all shown pictures, primary human osteoblasts were cultivated for three days under standard conditions. Reference material (**a**) 200 k and (**b**) 725 k (magnification). Cast specimens (**c**) 200 k and (**d**) 625 k (magnification). Cast specimens with ion beam implantation of Ca (ion density 1×10^{-16} cm^2, energy level 30 keV) ((**e**) 200 k and (**f**) 800 k magnification). Cast discs with ion beam implantation of P (ion density 1×10^{-16} cm^2, energy level 30 keV) ((**g**) 200 k and (**h**) 800 k magnification). Scanning electron microscopy was performed using a Hitachi S-5200 (Tokyo, Japan). Pictures were taken from [18] with permission.

It can be summarized that conspicuous changes in cell morphology could be detected on none of the investigated casting surfaces. Therefore, with regard to cell adhesion and cell spreading, an equivalence between centrifugal casting and the current reference surfaces can be assumed. Additionally, the ion beam implantation of calcium or phosphorus ions did not obviously affect the respective cellular response.

3.2. Cell Adhesion and Proliferation

In terms of biocompatibility, the cell adhesion and proliferation/viability measured by MTT-assay can be used as a well-established tool for characterizing cellular interaction with biomaterials. As shown in Figure 5, on all tested surfaces the human osteoblasts could adhere in equal amounts after 24 h. No statistically significant difference was observed between the reference material and the specimen produced by centrifugal casting. In addition, after cultivation for 7 days, no significant difference could be identified. Regarding the time of cultivation, a significant difference between day 1 and day 7 was detectable for the reference (REF) and the cast material (CAST). The ion implantation of P led to a nearly significant increase from day 1 to day 7 (CAST+P; $p = 0.055$). After the ion implantation of Ca, the mean cell number at day 7 was lowest compared with the other groups, and no significant increase compared with day 1 could be calculated ($p = 0.24$).

In summary, the type of production of the discs did not inhibit the adhesion and proliferation of human osteoblastic cells. All tested groups showed an increase in cell numbers over time.

In addition, by comparing all tested groups of material surfaces, no statistically significant difference was observed ($p > 0.99$).

3.3. Optimized Centrifugal Casting Did Not Impair Osteogenic Differentiation

In the presented work, we also investigated whether osteogenic differentiation of human osteoblasts was influenced by cultivation on cast surfaces or reference material. Target genes for the analysis of osteogenic differentiation markers on gene expression levels were RUNX2, SP7, COL1a1, ALP and BGLAP, and markers for bone homoeostasis were OPG, RANKL, and CASP3 for the induction of cell apoptosis, as shown in Figure 6.

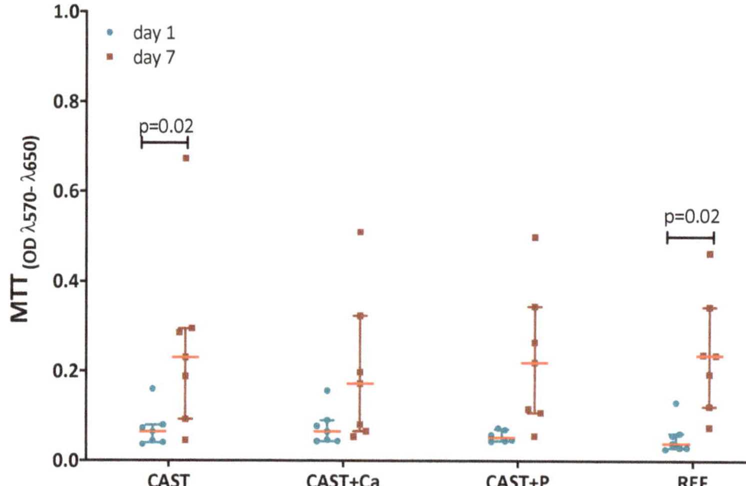

Figure 5. Cell proliferation tested by MTT-assay. Besides the comparison of reference material (REF) and cast specimens (CAST), the influence of beamline-implanted Ca- or P-ions into surfaces on cytotoxicity/proliferation was also tested. The proliferation of osteoblasts on the surfaces was investigated on day 1 and day 7 (n = 7 each). The quotients of OD 570–650 nm were measured, and respective values are shown as scatter plots with Median and interquartile ranges. Influences of time as well as surfaces were analyzed by two-way ANOVA and Bonferroni post hoc tests for statistical significance. The significance level was set to $p \leq 0.05$. OD: Optical Density; Ca: Ca-ion; P: P-ion; Implantation dose 1×10^{-16} cm^2, implantation energy 30 keV.

The Runt-related transcription factor 2 (RUNX2) as a key regulator of osteoblast differentiation was expressed in equal amounts on all tested surfaces. Additionally, the expression of transcription factor SP7 (Osterix), a regulator of osteogenic differentiation that enhances the effect of RUNX2, was not modulated by the different surfaces. So, the main transcriptional regulators of osteogenic differentiation were not influenced by the centrifugal cast materials.

In addition, later relevant gene expressions of the osteoblastic phenotype, like COL1a1 (collagen type I, the essential collagenous component for building bone matrix), ALP (Alkaline phosphatase, needed for mineralization of bone matrix), and BGLAP (Osteocalcin, needed to build hydroxyapatite crystals within the bone matrix), were not significantly modulated over the tested time period of 7 days.

With respect to bone resorption, the tested marker genes OPG (inhibitor of bone resorption) and RANKL (stimulator of bone resorption) showed no significant differences between the CAST and reference specimen (OPG $p = 0.22$, RANKL $p > 0.99$). However, the implantation of Ca- ($p = 0.015$) or P-ions ($p = 0.005$) significantly reduced the gene expression level for OPG. The expression for RANKL was not modified for any group ($p > 0.99$).

The induction of apoptosis to the osteoblasts was tested by CASP3 analysis. Caspase 3 is a key regulator for cell death. No significant difference ($p = 0.999$) on gene expression level was observed after 7 days of cultivation on the tested specimen.

In summary, on gene expression level, the spin-cast and the reference material did not show significant differences. In terms of cast material, the additional Ca-/P-ion implantation significantly reduced the gene expression level of OPG.

We further analyzed the ongoing differentiation process of protein and on mineralization levels by quantification of C-terminal collagen type I propeptide (CICP) as a measure of collagen type I synthesis and Osteoprotegerin (OPG), as well as Alizarin red S staining, as shown in Figure 7. All data on protein expression were normalized to the respective MTT values as a measure of cell number.

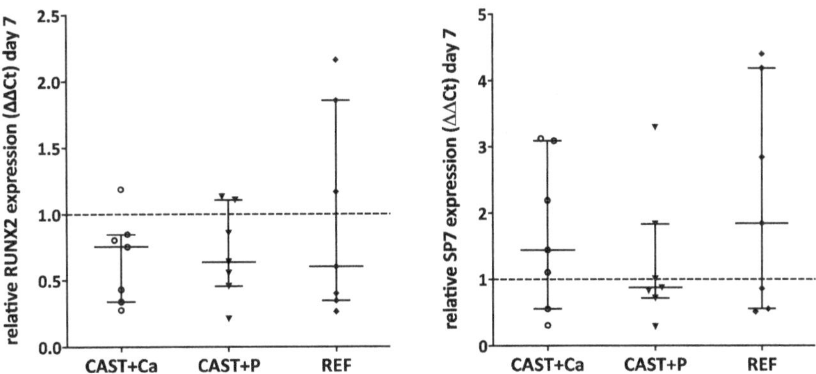

(**a**) Regulators of gene expression RUNX2 and SP7.

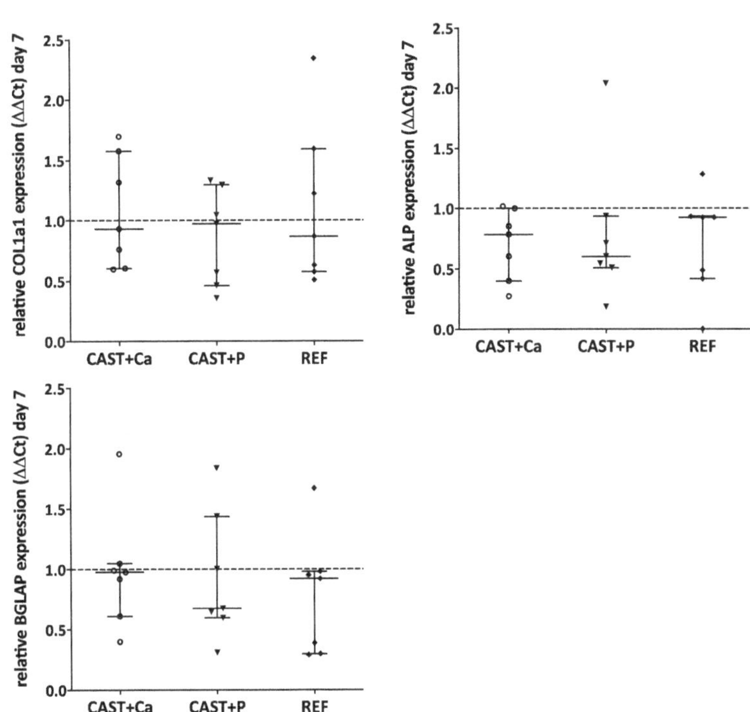

(**b**) Markers for differentiation COL1a1, ALP and BGLAP.

Figure 6. *Cont.*

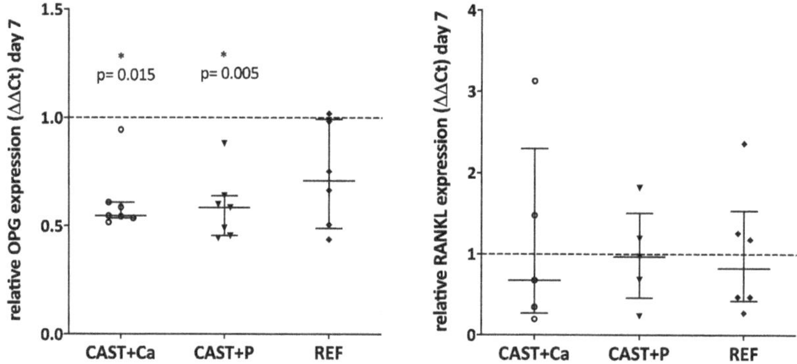

(**c**) Marker for bone homeostasis OPG and RANKL.

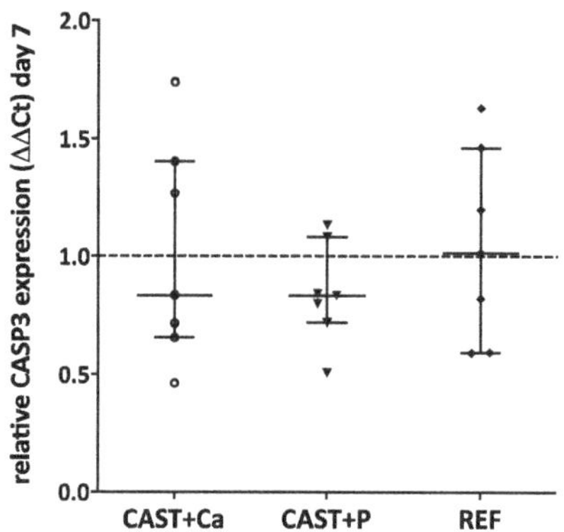

(**d**) Marker for apoptotic process CASP3.

Figure 6. Gene expression analysis of primary human osteoblasts after cultivation for 7 days on a cast Ti6Al4V specimen without and with Ca- or P-ion implantation as well as a reference Ti6Al4V-specimen. Diagramed are scatterplots showing the interquartile range of the calculated ΔΔCt gene expression. (**a**) Shows results of regulators of the osteogenic phenotype. While RUNX2 (Runt-related transcription factor 2) is a key regulator in osteoblast differentiation, SP7 (Osterix) is a regulator of osteogenic differentiation and enhances the effect of RUNX2. (**b**) Presents the gene expression of differentiated osteoblasts. COL1a1: collagen type I is the major structural protein of bone. ALP: alkaline phosphatase is needed for mineralization of bone matrix and BGLAP: osteocalcin is needed to build hydroxyapatite crystals within the bone matrix. (**c**) Markers for bone homeostasis. OPG: osteoprotegerin is a decoy receptor for RANKL (receptor activator of nuclear factor kappa-B ligand), while RANKL normally binds to RANK in order to stimulate bone resorption by osteoclasts. (**d**) Marker for the apoptotic process. CASP3: caspase 3 is a predominant caspase and regulates cell apoptosis. Gene expression was normalized in osteoblasts cultivated on cast surfaces without ion implantation. All gene expression data were normalized internal to HPRT1 as a housekeeping gene. Statistical analysis by one-way ANOVA with Kruskal–Wallis and Dunn's post hoc test. The significance level was set to $p < 0.05$ (*). Ca: Ca-ion; P: P-ion. Ion dose 1×10^{-16} cm^2, implantation energy 30 keV; $n = 7$.

The CICP concentration in the cell culture supernatant, reflecting the amount of procollagen type I biosynthesis by the osteoblasts seeded on cast surfaces, declined from day 3 (730.2 pg/mL) until day 7 (272.4 pg/mL, $p < 0.07$), indicating a negative feedback mechanism of pericellular collagen type I accumulation (Figure 7a). Osteoblasts seeded on reference surfaces produced, according to the current standard method, similar results (day 3: 912 pg/mL; day 7: 240 pg/mL) for CICP-concentration, although the decline was statistically significant in this case. In the cell culture supernatant of Ca-ion implanted casting surfaces, a CICP level of 706.8 pg/mL could be observed, while on day 7 the secretion decreased 1.7-fold to 415 pg/mL ($p > 0.99$). The amount of CICP in the cell culture supernatant of P-ion implanted casting surfaces was somewhat lower compared with Ca-ion implanted surfaces, although the difference was not significant. Over time, the median CICP of 611.5 pg/mL decreased by a factor of 2.5 to 244 pg/mL ($p = 0.09$) on day 7. In the cell culture supernatant of the reference surfaces, a 3.8-fold significant decrease in CICP to 239.5 pg/mL was observed on day 7 ($p = 0.005$). Overall, no significant difference in type I procollagen production of osteoblasts on cast surfaces without or with ion implantation could be determined compared with the reference surfaces.

OPG served as a bone turnover marker and is produced by osteoblasts to counteract bone resorption by osteoclasts. OPG is a decoy receptor for RANKL in the RANK/RANKL/OPG axis and is essential to bone metabolism. On the non-ion-implanted casting surfaces, normalized to MTT-values, an OPG release of 2673 pg/mL could be measured on day 7 (Figure 7b). Furthermore, it was shown that the OPG concentration in the cell culture supernatant of Ca-ion implanted cast surfaces was 3840 pg/mL and thus 1.4 times higher than P-ions implanted surfaces (2779 pg/mL). The OPG secretion of the osteoblasts, cultivated on reference surfaces, was similar to cast surfaces without and with P-ion implanted surfaces (median of 2833 pg/mL). Overall, there were no significant differences for OPG release of human osteoblastic cells on the surfaces ($p < 0.99$).

Figure 7c presents data of an Alizarin red S staining of osteoblast-like cells cultivated on cast material and on reference material shown after 14 days. The median Alizarin red S concentration of osteoblasts on casting surfaces was 139.0 µg/mL. For reference discs, a median Alizarin red S concentration of 152.3 µg/mL was measured. There was no significant difference between reference and cast surfaces ($p = 0.88$). The Alizarin red S values of the cells cultivated on the Ca-ion implanted surfaces was significantly reduced by 27.8% compared with non-modified cast surfaces ($p = 0.02$). For P-ion implantation in casting material, the values were also somewhat lower compared with non-ion-implanted material but not significantly different ($p = 0.16$). Neither comparing CAST-Ca ($p = 0.15$) nor CAST-P ($p = 0.30$) with the reference material showed statistically significant differences.

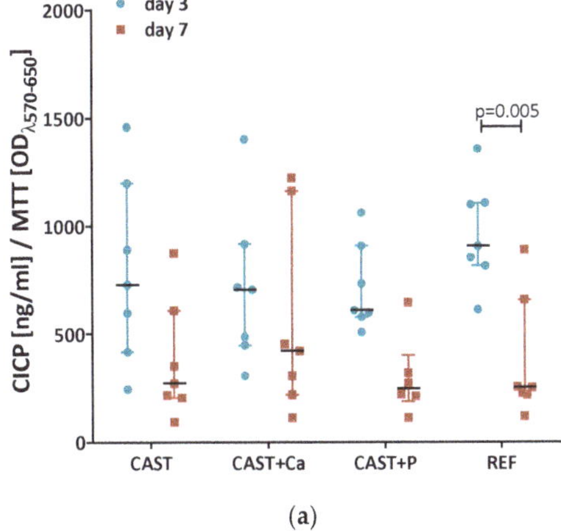

(a)

Figure 7. *Cont.*

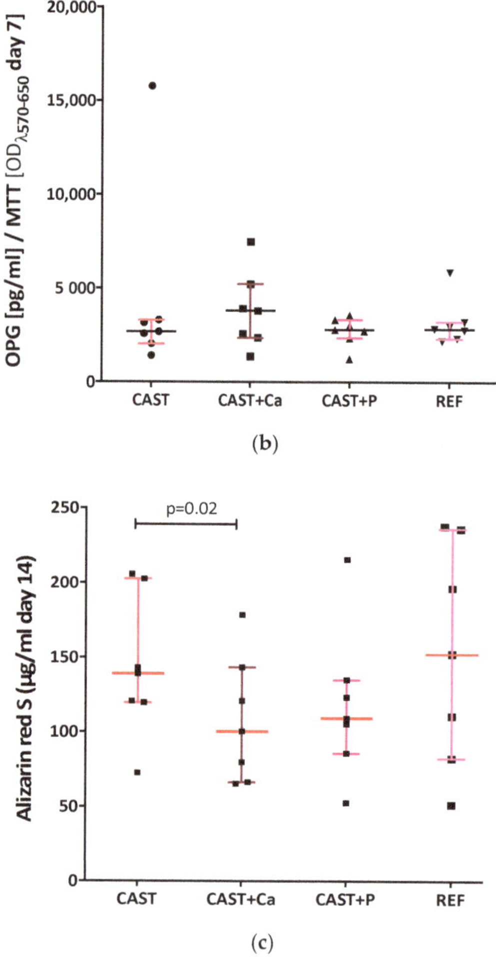

Figure 7. Detection of in vitro mineralization and protein secretion. (**a**) CICP Elisa for collagen type I synthesis after osteoblast cultivation for 3 and 7 days. Influence of different surfaces on collagen type I synthesis was tested by a two-way ANOVA with Bonferroni post hoc test. (**b**) OPG Elisa results of osteoblast cultures after 7 days cultivation. Data were analyzed by using Kruskal–Wallis and Dunn's post hoc tests for statistical significance. (**c**) Alizarin red S staining was used to analyze the mineralization of osteoblasts' seeded specimens and the effect of Ca- or P-Ion coated surfaces after 14 days. Statistical analysis was done by a one-way ANOVA with a Tukey post hoc test. The significance level was set to $p \leq 0.05$. Values are shown as scatter plots with Median and interquartile range. Ca-ion and P-ion implantation by ion dose 1×10^{-16} cm^2, implantation energy 30 keV; $n = 7$.

In summary, this cell biologic analyses showed that there was no significant difference between the examined cast surfaces and the reference surfaces with regard to procollagen type I and OPG secretion as well as in vitro mineralization. The additional ion implantation of Ca or P did not show any supportive effect either.

4. Discussion

We aimed to show that an optimized spin casting process of Ti6Al4V, a frequently used titanium alloy for the production of hip endoprostheses for uncemented implantation, can generate material surface characteristics with comparable in vitro biocompatibility for primary human osteoblasts as the standard machined manufacturing process. Ion beam implantation of calcium or phosphorus following aluminum oxide blasting of the cast material, however, did not further improve the cell-biologic outcome in the dosage used.

With respect to material properties, standardized analysis did not reveal an indication of Alpha Case formation, confirming the positive effect of the optimized casting process with respect to metallurgical quality. Furthermore, chemical composition of the cast parts was within the specification. Therefore, no relevant reaction between melt and crucible or shell mold could be observed. Additionally, subsequent aluminum oxide blasting led to expected surface roughness with no significant difference to the reference machined material.

In vivo bone response to cast titanium implants in the tibia metaphysis has been previously studied in a rabbit and a rat model [4,19]. Mohammadi et al. used cast pure titanium from which about 0.25 mm of the superficial layer was eliminated by machining to remove impurities created by the casting process [19]. Our recent publication on a rat model was based on the same optimized casting process of the titanium alloy Ti6Al4V without removal of the superficial layer before aluminum oxide blasting, as used in the present study [4]. In both animal models, no significant difference in osseointegration after 3 months (rabbit and rat) or 6 months (rabbit) in comparison to the respective machined control implants could be observed. Recently, with respect to the treatment of large bone defects, a porous cast titanium alloy (Ti6Al7Nb) was reported to possess good in vitro and in vivo biocompatibility after acid etching [20]. We are not aware, however, of any previous in vitro studies on the cell-biologic response of osteogenic cells in contact with cast titanium or titanium alloy without mechanical or chemical removal of the superficial layer.

To address this knowledge gap in a clinically relevant manner, we used the established optimized casting process for Ti6Al4V, an aluminum oxide blasting process currently included in the commercial medical implant production and tested the interaction of primary human osteoblasts from patients with end-stage knee osteoarthritis who received joint-replacement surgery as a cell source. Moreover, a broad set of relevant central cell-biologic outcome parameters, such as osteoblast adhesion and proliferation, markers of early and late osteogenic differentiation, as well as secretion of signaling molecules reflecting bone turnover, was analyzed.

For both reference and cast material, we observed relatively large variation in the cell-biologic outcome parameters. This can be attributed to the use of primary human osteoblast cultures from different patients and is a common observation for non-cell lines. Nevertheless, testing primary human cells instead of using commercially available (immortalized or tumor-derived) cell-lines has been recommended in the literature [21].

Regarding cell adhesion and the proliferation of human osteoblasts on the tested specimen of spin cast or machined Ti6Al4V material, no difference was detectable. For the interpretation of these results, it is important that the process of aluminum oxide blasting resulted in nearly identical surface roughness. That the microstructure of the surface represents an essential factor for cell adhesion, proliferation and differentiation is well known and has been the subject of comprehensive reviews, for example by Mitra et al. [22]. Yokose and co-worker demonstrated that the surface micro topography of titanium discs influences osteoblast-like cell proliferation and differentiation [23]. Moreover, Hatano et al. reported that primary rat osteoblasts showed higher proliferation, and alkaline phosphatase and osteocalcin expression cultivated on rough rather than smooth tissue culture polystyrene [24]. It is also known from the literature that roughness of the surface is important for a successful osseointegration [25]. In addition to classical micro and nano topography, the presence of pores within the surface also has a major influence on cell adhesion and differentiation of the desired cells [22]. Thus, further improvement of integration can be achieved, for example, by using multi scale porous titanium surfaces rather than smooth or micro-roughened titanium [25]. According to results from Anselme and coworkers, our findings indicate a similar biocompatibility of the reference and cast material since the surface topography, generated by classical aluminum-oxide blasting, was not different [26,27]. As previously reported by Lagonegro et al., we observed that human osteoblasts preferentially adhered to the peaks of micro-topography and bridged over geometric surface irregularities [23,28].

A similar observation was published by Yin et al. [29]. The authors showed that the blasting or etching of a titanium surface has a significant effect on osteoblast differentiation. Rough-blasted surfaces supported the differentiation process while the etching process reduced the expression of osteoblast markers like RUNX2, COL1a1, and ALP. We also tested these osteogenic differentiation markers on human primary osteoblasts and found no significant difference between the reference machined material and the cast specimen. The same result was obtained for the expression of BGLAP as a late marker of osteoblast phenotype and the in vitro mineralization determined by Alizarin Red Assay. These findings are in line with the recently published results of Wölfle-Roos et al. [4]. They showed that in rats in vivo no negative effect with respect to bone-implant contact or pull-out-force could be attributed to the optimized manufacturing process, which was also used in the present in vitro experiments.

Some titanium surfaces in the in vitro study were further modified by ion beam implantation of Ca- or P-ions. Comparing our data with the available literature on similar surface modifications, some divergences need to be discussed. Krupa et al. presented similar results for human bone derived cells when seeded on mirror polished pure commercial Ti-surfaces. They used similar conditions for ion implantation and found no adverse biological effects in terms of cell viability and ALP analysis [30,31]. In addition, Nayab et al. could not detect any effect on MG63 cells after seeding on pure Ti-surfaces with implanted Ca ions (ion dose 1×10^{-15} or 1×10^{-16}). Only by using higher concentrations of Ca (ion dose 1×10^{-17}) did they describe slightly better cell spreading, associated with delayed adhesion and enhanced expression of bone cell markers [32,33]. Several working groups could demonstrate a positive effect on cellular behavior by coating titanium surfaces with calcium or phosphor [34–37]. Besides the different metallic materials, pure titanium vs. Ti6Al4V, they also used different techniques for surface modification like plasma spraying or chemical methods. Therefore, the amount of biologically available ions is not comparable [33]. By using ion implantation, the ions become dispersed into a certain range of depth and a relevant part of them becomes not bioavailable [13]. Thus, besides ion density, implantation energy also influences biological impact. Moreover, the vast majority of in vitro studies are not based on sand-blasted or etched surfaces with a roughness comparable to the current clinical application. In the case of a relevant surface micro-roughness plasma, ion implantation of calcium, used in the study of Cheng et al., could also have advantages [38]. The authors found significant positive effects of calcium plasma immersion ion implantation on the osteoblast-like cell line MG63 with respect to adhesion, proliferation and osteogenic differentiation in vitro and in a rabbit in vivo model. However, besides the different ion implantation technique and dosage, the use of pure machined titanium, a lower degree of surface micro-roughness and the analysis of a bone-tumor derived cell-line instead of primary human osteoblasts in this study precludes a direct comparison. The encouraging results nevertheless indicate that different strategies and dosages of ion implantation should be further studied with our optimized spin cast Ti6Al4V-alloy.

Additionally, the charge of the surface modulates cellular activities. In our study, the addition of Ca- or P-ions by ion beam implantation even negatively affected mineral deposition in the synthesized collagen matrix, as shown by Alizarin red S staining. This is somewhat contrary to the so far known literature [36,39]. One explanation could be the fact that these groups used different methods for surface modifications than we did. They used chemical approaches or, like Knabe et al., plasma spraying to cover the titanium with Ca- or Zn-P-ions [35]. It could be possible that the surface charge was not changed in an analogous manner in addition to a different amount of deposition and the bioavailability of the ions when using these alternative surface modification techniques. In future experiments, a similar approach to investigating surface charge changes as described by Kulkarni and co-workers could be used [40]. They presented data on how surface characteristics influence interaction with small sized model proteins. Namely, variations in charge densities towards the top edges of the surface seem to be a relevant factor. This effect was shown to be mainly triggered by nano topography [41]. Due the identical production process of our cast specimen and the not significantly different results of Ra and Rz without/with ion implantation, we conclude that any possible differences in the charge density in the case of ion implantation could be due to the deposition itself or to influences on the nanostructure, which was not analyzed in this study. Future experiments should consider studying nano-topographic features and analysis of wettability as well as zeta potential.

In terms of cell apoptosis, we tested for the gene expression of CASP3, a key member of the apoptotic pathway in osteogenic cells undergoing differentiation [42]. Even using the same alloy for reference and cast specimens, treating them in the same medical grade production process, including aluminum oxide blasting and sterilization procedures with the same machinery, it could have been possible that different micro- or nano-topographic structures could have led to unwanted cell death of the osteoblasts. This relevant topic has been disused by several groups. Kulkarni et al. showed that the titanium alloy surface topography influences cell survival and cell death [43]. Unfortunately, the rough micro-topography of the specimen did not allow for a reliable measurement of the nano-topography that could somehow be different in the machined references and the cast specimen. Besides that, some knowledge arises that cytotoxic effects of Titanium and its alloys have to be considered [44]. Nevertheless, Ti6Al4V is the most frequently and successfully used alloy in orthopedic surgery because of its biocompatibility and certain bone-similar characteristics. Therefore, the present work focused on the transfer of the routine fabrication of this alloy to a newly developed fabrication method.

It was also reported that, besides the overexpression of CASP3, which leads to unwanted cell death, a loss of CASP3 expression results in delayed ossification and decreased bone mineral density. Comparing our reference material with that of the cast, we could conclude that the given CASP3 expression does not impair the osteogenic differentiation process and preserves the viability of osteoblasts. Also, our data indicate no respective difference in CASP3 expression between the tested specimens.

Because of the non-loaded in vitro conditions, a possible impact on inflammatory processes should be further tested by including, for example, abrasion experiments and by studying interaction with macrophages. Under unloaded in vitro conditions, the analysis of osteoblast OPG and RANKL expression on the gene and protein level did not indicate any disadvantage of the optimized casting process in terms of secretion of central regulators of osteoclast activity in the present study. These results are in line with our data from the in vivo study in which no relevant induction of inflammation or subsequent bone resorption or local osteolysis was observed after the implantation of small rods in the tibia metaphysis of rats [4]. Nevertheless, it should be kept in mind that further testing in a load-bearing large animal model still has to be performed.

The optimized casting process was successfully developed in terms of methods and practical procedure and is currently the subject of further research, aiming to create a scalable production process for market-ready implants. So far, produced tibia and femur prototypes meet given demands regarding microstructure as well as mechanical and machining properties. Approaching industrial part production, the process is now in the development stage to facilitate upscaled production lots for the fabrication of uncemented endoprostheses at competitive market prices. Besides the still-required increase in productivity, the improvement of dimensional accuracy is being addressed as a secondary challenge as methods to correct it are successfully employed regularly. Finally, in addition to comparison with standard machined Ti6Al4V, material comparative testing of spin cast Ti6Al4V with other Titanium alloys, like Ti6Al7Nb or Ti28NB35.4Zr, should be performed.

5. Conclusions

It can be stated that the surfaces produced in the optimized casting process of Ti6Al4V are not different from the reference surfaces of the same alloy successfully used in orthopedic surgery with respect to cell-biologic interactions of human osteoblasts in vitro. Neither a difference in cell spreading, cell viability/proliferation nor an influence on osteogenic differentiation and mineralization of primary osteoblasts was detectable. The optimized casting process, therefore, provided equivalent biologic material quality in this first comprehensive in vitro analysis based on primary human osteoblasts. These results confirm our recent in vivo data on osseointegration in a rat model. [4] Thus, near net-shape precision casting of Ti6Al4V, avoiding high demand of stock material for machining as well as the cost for the machining itself, may present a promising and cost-effective alternative to the conventional machining-based process for uncemented orthopedic implants.

Ion implantation of calcium or phosphorus by ion beam, however, did not induce a positive effect on human osteoblastic cells in vitro, again confirming our previous data on osseointegration in the rat model [4] despite the use of different ion-implantation techniques (ion-beam versus plasma ion implantation), which makes the results not directly comparable. Therefore, further studies on the effect of different ion implantation doses should be performed.

Author Contributions: Conceptualization, B.R.E., F.J., M.H., K.G. and K.A.B.; methodology, B.R.E., F.J., M.H., K.G. and K.A.B.; validation, B.R.E., F.J., M.H., K.G. and K.A.B.; formal analysis, K.A.B., M.H. and K.G.; investigation, K.A.B., M.H. and K.G.; resources, B.R.E., M.H. and K.G.; data curation, F.J., M.H. and K.G.; writing—original draft preparation, J.F.; writing—review and editing, F.J., B.R.E., M.H., K.A.B. and K.G.; visualization, F.J. and M.H.; supervision, B.R.E.; project administration, B.R.E., M.H., K.G. and K.A.B.; funding acquisition, B.R.E. All authors have read and agreed to the published version of the manuscript.

Funding: This study has been funded by a grant from the German Federal Ministry of Education and Research (BMBF, No. 13 GW0020B/E).

Acknowledgments: We wish to acknowledge the support of Giovanni Ravalli for technical assistance and Paul Walther (Department of electron microscopy, University of Ulm, Germany) for his support in the electron microscopic studies). We gratefully thank Andreas Kolitsch from the Helmholtz-Zentrum Dresden-Rosssendorf (HZDR) for the ion beam implantation.

Conflicts of Interest: The funders had no role in the design of the study; in the collection, analyses, or interpretation of data; in the writing of the manuscript, or in the decision to publish the results. The authors declare no conflict of interest.

References

1. Ha, S.-W.; Wintermantel, E. Biokompatible Metalle. In *Medizintechnik*, 5th ed.; Wintermantel, E., Ha, S., Eds.; Springer: Berlin/Heidelberg, Germany, 2009; pp. 191–217. [CrossRef]
2. Nastac, L.; Gungor, M.N.; Ucok, I.; Klug, K.L.; Tack, W.T. Advances in investment casting of Ti–6Al–4V alloy: A review. *Int. J. Cast Met. Res.* **2013**, *19*, 73–93. [CrossRef]
3. Billhofer, H.; Hauptmann, T. Feingusssystem für Titan und Titanlegierungen. *Lightweight Des.* **2010**, *3*, 47–51. [CrossRef]
4. Wolfle-Roos, J.V.; Katmer Amet, B.; Fiedler, J.; Michels, H.; Kappelt, G.; Ignatius, A.; Durselen, L.; Reichel, H.; Brenner, R.E. Optimizing Manufacturing and Osseointegration of Ti6Al4V Implants through Precision Casting and Calcium and Phosphorus Ion Implantation? In Vivo Results of a Large-Scale Animal Trial. *Materials* **2020**, *13*, 1670. [CrossRef]
5. Kienapfel, H.; Sprey, C.; Wilke, A.; Griss, P. Implant fixation by bone ingrowth. *J. Arthroplast.* **1999**, *14*, 355–368. [CrossRef]
6. Wolfle, J.V.; Fiedler, J.; Durselen, L.; Reichert, J.; Scharnweber, D.; Forster, A.; Schwenzer, B.; Reichel, H.; Ignatius, A.; Brenner, R.E. Improved anchorage of Ti6Al4V orthopaedic bone implants through oligonucleotide mediated immobilization of BMP-2 in osteoporotic rats. *PLoS ONE* **2014**, *9*, e86151. [CrossRef]
7. Krupa, D.; Baszkiewicz, J.; Kozubowski, J.A.; Barcz, A.; Sobczak, J.W.; Bilinski, A.; Lewandowska-Szumiel, M.; Rajchel, B. Effect of dual ion implantation of calcium and phosphorus on the properties of titanium. *Biomaterials* **2005**, *26*, 2847–2856. [CrossRef]
8. Lazarinis, S.; Makela, K.T.; Eskelinen, A.; Havelin, L.; Hallan, G.; Overgaard, S.; Pedersen, A.B.; Karrholm, J.; Hailer, N.P. Does hydroxyapatite coating of uncemented cups improve long-term survival? An analysis of 28,605 primary total hip arthroplasty procedures from the Nordic Arthroplasty Register Association (NARA). *Osteoarthr. Cartil.* **2017**, *25*, 1980–1987. [CrossRef]

9. Surmenev, R.A.; Surmeneva, M.A.; Ivanova, A.A. Significance of calcium phosphate coatings for the enhancement of new bone osteogenesis—A review. *Acta Biomater.* **2014**, *10*, 557–579. [CrossRef]
10. Pham, M.T.; Reuther, H.; Matz, W.; Mueller, R.; Steiner, G.; Oswald, S.; Zyganov, I. Surface induced reactivity for titanium by ion implantation. *J. Mater. Sci. Mater. Med.* **2000**, *11*, 383–391. [CrossRef]
11. Rautray, T.R.; Narayanan, R.; Kwon, T.Y.; Kim, K.H. Surface modification of titanium and titanium alloys by ion implantation. *J. Biomed. Mater. Res. B Appl. Biomater.* **2010**, *93*, 581–591. [CrossRef]
12. Biggs, M.J.; Richards, R.G.; Gadegaard, N.; Wilkinson, C.D.; Dalby, M.J. The effects of nanoscale pits on primary human osteoblast adhesion formation and cellular spreading. *J. Mater. Sci. Mater. Med.* **2007**, *18*, 399–404. [CrossRef] [PubMed]
13. Dalby, M.J.; Yarwood, S.J.; Riehle, M.O.; Johnstone, H.J.; Affrossman, S.; Curtis, A.S. Increasing fibroblast response to materials using nanotopography: Morphological and genetic measurements of cell response to 13-nm-high polymer demixed islands. *Exp. Cell Res.* **2002**, *276*, 1–9. [CrossRef] [PubMed]
14. Kiang, J.D.; Wen, J.H.; del Alamo, J.C.; Engler, A.J. Dynamic and reversible surface topography influences cell morphology. *J. Biomed. Mater. Res. A* **2013**, *101*, 2313–2321. [CrossRef]
15. Fiedler, J.; Kolitsch, A.; Kleffner, B.; Henke, D.; Stenger, S.; Brenner, R.E. Copper and silver ion implantation of aluminium oxide-blasted titanium surfaces: Proliferative response of osteoblasts and antibacterial effects. *Int. J. Artif. Organs* **2011**, *34*, 882–888. [CrossRef] [PubMed]
16. Fiedler, J.; Ozdemir, B.; Bartholoma, J.; Plettl, A.; Brenner, R.E.; Ziemann, P. The effect of substrate surface nanotopography on the behavior of multipotnent mesenchymal stromal cells and osteoblasts. *Biomaterials* **2013**, *34*, 8851–8859. [CrossRef]
17. Mane, V.P.; Heuer, M.A.; Hillyer, P.; Navarro, M.B.; Rabin, R.L. Systematic method for determining an ideal housekeeping gene for real-time PCR analysis. *J. Biomol. Tech.* **2008**, *19*, 342–347.
18. Katmer Amet, B. Bewertung von Ti-6Al-4V Oberflächen Hergestellt Mittels neu Entwickelter Schleudergusstechnologie im In Vitro- und In Vivo-Modell. Ph.D. Thesis, University of Ulm, Ulm, Germany, 2019. [CrossRef]
19. Mohammadi, S.; Esposito, M.; Wictorin, L.; Aronsson, B.O.; Thomsen, P. Bone response to machined cast titanium implants. *J. Mater. Sci.* **2001**, *36*, 1987–1993. [CrossRef]
20. Sommer, U.; Laurich, S.; de Azevedo, L.; Viehoff, K.; Wenisch, S.; Thormann, U.; Alt, V.; Heiss, C.; Schnettler, R. In Vitro and In Vivo Biocompatibility Studies of a Cast and Coated Titanium Alloy. *Molecules* **2020**, *25*, 3399. [CrossRef]
21. Czekanska, E.M.; Stoddart, M.J.; Ralphs, J.R.; Richards, R.G.; Hayes, J.S. A phenotypic comparison of osteoblast cell lines versus human primary osteoblasts for biomaterials testing. *J. Biomed. Mater. Res. A* **2014**, *102*, 2636–2643. [CrossRef]
22. Mitra, J.; Tripathi, G.; Sharma, A.; Basu, B. Scaffolds for bone tissue engineering: Role of surface patterning on osteoblast response. *RSC Adv.* **2013**, *3*, 11073–11094. [CrossRef]
23. Yokose, S.; Klokkevold, P.R.; Takei, H.H.; Kadokura, H.; Kikui, T.; Hibino, Y.; Shigeta, H.; Nakajima, H.; Kawazu, H. Effects of surface microtopography of titanium disks on cell proliferation and differentiation of osteoblast-like cells isolated from rat calvariae. *Dent. Mater. J.* **2018**, *37*, 272–277. [CrossRef] [PubMed]
24. Hatano, K.; Inoue, H.; Kojo, T.; Matsunaga, T.; Tsujisawa, T.; Uchiyama, C.; Uchida, Y. Effect of surface roughness on proliferation and alkaline phosphatase expression of rat calvarial cells cultured on polystyrene. *Bone* **1999**, *25*, 439–445. [CrossRef]
25. Jang, T.S.; Jung, H.D.; Kim, S.; Moon, B.S.; Baek, J.; Park, C.; Song, J.; Kim, H.E. Multiscale porous titanium surfaces via a two-step etching process for improved mechanical and biological performance. *Biomed. Mater.* **2017**, *12*, 025008. [CrossRef] [PubMed]
26. Anselme, K. Osteoblast adhesion on biomaterials. *Biomaterials* **2000**, *21*, 667–681. [CrossRef]
27. Anselme, K.; Noel, B.; Hardouin, P. Human osteoblast adhesion on titanium alloy, stainless steel, glass and plastic substrates with same surface topography. *J. Mater. Sci. Mater. Med.* **1999**, *10*, 815–819. [CrossRef]
28. Lagonegro, P.; Trevisi, G.; Nasi, L.; Parisi, L.; Manfredi, E.; Lumetti, S.; Rossi, F.; Macaluso, G.M.; Salviati, G.; Galli, C. Osteoblasts preferentially adhere to peaks on micro-structured titanium. *Dent. Mater. J.* **2018**, *37*, 278–285. [CrossRef]
29. Yin, C.; Zhang, Y.; Cai, Q.; Li, B.; Yang, H.; Wang, H.; Qi, H.; Zhou, Y.; Meng, W. Effects of the micro-nano surface topography of titanium alloy on the biological responses of osteoblast. *J. Biomed. Mater. Res. A* **2017**, *105*, 757–769. [CrossRef]

30. Krupa, D.; Baszkiewicz, J.; Kozubowski, J.A.; Barcz, A.; Sobczak, J.W.; Biliniski, A.; Lewandowska-Szumiel, M.D.; Rajchel, B. Effect of calcium-ion implantation on the corrosion resistance and biocompatibility of titanium. *Biomaterials* **2001**, *22*, 2139–2151. [CrossRef]
31. Krupa, D.; Baszkiewicz, J.; Kozubowski, J.A.; Barcz, A.; Sobczak, J.W.; Bilinski, A.; Lewandowska-Szumiel, M.; Rajchel, B. Effect of phosphorus-ion implantation on the corrosion resistance and biocompatibility of titanium. *Biomaterials* **2002**, *23*, 3329–3340. [CrossRef]
32. Nayab, S.N.; Jones, F.H.; Olsen, I. Effects of calcium ion implantation on human bone cell interaction with titanium. *Biomaterials* **2005**, *26*, 4717–4727. [CrossRef]
33. Nayab, S.N.; Jones, F.H.; Olsen, I. Effects of calcium ion-implantation of titanium on bone cell function in vitro. *J. Biomed. Mater. Res. A* **2007**, *83*, 296–302. [CrossRef] [PubMed]
34. Hotchkiss, K.M.; Reddy, G.B.; Hyzy, S.L.; Schwartz, Z.; Boyan, B.D.; Olivares-Navarrete, R. Titanium surface characteristics, including topography and wettability, alter macrophage activation. *Acta Biomater.* **2016**, *31*, 425–434. [CrossRef] [PubMed]
35. Knabe, C.; Berger, G.; Gildenhaar, R.; Klar, F.; Zreiqat, H. The modulation of osteogenesis in vitro by calcium titanium phosphate coatings. *Biomaterials* **2004**, *25*, 4911–4919. [CrossRef] [PubMed]
36. Kwon, Y.S.; Park, J.W. Osteogenic differentiation of mesenchymal stem cells modulated by a chemically modified super-hydrophilic titanium implant surface. *J. Biomater. Appl.* **2018**, *33*, 205–215. [CrossRef]
37. Vilardell, A.M.; Cinca, N.; Garcia-Giralt, N.; Dosta, S.; Cano, I.G.; Nogues, X.; Guilemany, J.M. Osteoblastic cell response on high-rough titanium coatings by cold spray. *J. Mater. Sci. Mater. Med.* **2018**, *29*, 19. [CrossRef]
38. Cheng, M.; Qiao, Y.; Wang, Q.; Jin, G.; Qin, H.; Zhao, Y.; Peng, X.; Zhang, X.; Liu, X. Calcium Plasma Implanted Titanium Surface with Hierarchical Microstructure for Improving the Bone Formation. *ACS Appl. Mater. Interfaces* **2015**, *7*, 13053–13061. [CrossRef]
39. Park, J.W.; Hanawa, T.; Chung, J.H. The relative effects of Ca and Mg ions on MSC osteogenesis in the surface modification of microrough Ti implants. *Int. J. Nanomed.* **2019**, *14*, 5697–5711. [CrossRef]
40. Kulkarni, M.; Mazare, A.; Park, J.; Gongadze, E.; Killian, M.S.; Kralj, S.; von der Mark, K.; Iglic, A.; Schmuki, P. Protein interactions with layers of TiO2 nanotube and nanopore arrays: Morphology and surface charge influence. *Acta Biomater.* **2016**, *45*, 357–366. [CrossRef]
41. Gongadze, E.; Kabaso, D.; Bauer, S.; Slivnik, T.; Schmuki, P.; van Rienen, U.; Iglic, A. Adhesion of osteoblasts to a nanorough titanium implant surface. *Int. J. Nanomed.* **2011**, *6*, 1801–1816. [CrossRef]
42. Miura, M.; Chen, X.D.; Allen, M.R.; Bi, Y.; Gronthos, S.; Seo, B.M.; Lakhani, S.; Flavell, R.A.; Feng, X.H.; Robey, P.G.; et al. A crucial role of caspase-3 in osteogenic differentiation of bone marrow stromal stem cells. *J. Clin. Investig.* **2004**, *114*, 1704–1713. [CrossRef]
43. Kulkarni, M.; Mazare, A.; Gongadze, E.; Perutkova, Š.; Kralj-Iglič, V.; Milošev, I.; Schmuki, P.; Iglič, A.; Mozetič, M. Titanium nanostructures for biomedical applications. *Nanotechnology* **2015**, *26*, 062002. [CrossRef] [PubMed]
44. Kim, K.T.; Eo, M.Y.; Nguyen, T.T.H.; Kim, S.M. General review of titanium toxicity. *Int. J. Implant Dent.* **2019**, *5*, 10. [CrossRef] [PubMed]

Publisher's Note: MDPI stays neutral with regard to jurisdictional claims in published maps and institutional affiliations.

Article

In Situ Synchrotron X-ray Diffraction Investigations of the Nonlinear Deformation Behavior of a Low Modulus *β*-Type Ti36Nb5Zr Alloy

Qingkun Meng [1], Huan Li [1], Kai Wang [1], Shun Guo [2,*], Fuxiang Wei [1], Jiqiu Qi [1,*], Yanwei Sui [1], Baolong Shen [1] and Xinqing Zhao [3,*]

1 School of Materials Science and Physics, China University of Mining and Technology, Xuzhou 221116, China; qkmeng@cumt.edu.cn (Q.M.); lihuan08162@163.com (H.L.); wk12037373@163.com (K.W.); 4637@cumt.edu.cn (F.W.); suiyanwei@cumt.edu.cn (Y.S.); shenbaolong@cumt.edu.cn (B.S.)
2 School of Materials Science and Engineering, Jiangsu University, Zhenjiang 212013, China
3 School of Materials Science and Engineering, Beihang University, Beijing 100191, China
* Correspondence: shunguo@ujs.edu.cn (S.G.); qijiqiu@cumt.edu.cn (J.Q.); xinqing@buaa.edu.cn (X.Z.); Tel.: +86-511-8878-3268 (S.G.); +86-516-8359-1879 (J.Q.); +86-10-8233-8559 (X.Z.)

Received: 4 November 2020; Accepted: 30 November 2020; Published: 2 December 2020

Abstract: The low modulus β-type Ti alloys usually have peculiar deformation behaviors due to their low phase stability. However, the study of the underlying mechanisms is challenging since some physical mechanisms are fully reversible after the release of the load. In this paper, the deformation behavior of a low modulus β-type Ti36Nb5Zr alloy was investigated with the aid of in situ synchrotron X-ray diffraction (SXRD) during tensile loading. The evolution of lattice strains and relative integrated diffraction peak intensities of both the β and α'' phases were analyzed to determine the characteristics of the potential deformation mechanisms. Upon loading, the α'' diffraction spots appeared at specific azimuth angles of the two-dimensional SXRD patterns due to the <110> fiber texture of original β grains and the selection of favorable martensitic variants. The nonlinear deformation behavior originated from a reversible stress-induced martensitic transformation (SIMT). However, the SIMT contributed a little to the large recoverable strain of over 2.0%, which was dominated by the elastic deformation of the β phase. Various deformation mechanisms were activated successively at different applied strains, including elastic deformation, SIMT and plastic deformation. Our investigations provide in-depth understandings of the deformation mechanisms in β-type Ti alloys with low elastic modulus.

Keywords: Ti alloys; martensitic transformation; recoverable strain; synchrotron X-ray diffraction

---

1. Introduction

Titanium (Ti) and its alloys been used extensively for biomedical applications due to their excellent combined properties of low elastic modulus, high specific strength, excellent corrosion resistance, complete inertness to body environment and superior biocompatibility [1,2]. Among the mechanical properties essential for implant materials, elastic modulus, whose value should be as close as possible to that of human bone, is of considerable importance [3]. Although the elastic modulus of the widely used pure Ti and Ti-6Al-4V is lower (104 GPa and 110 GPa, respectively) than that of other conventional metallic biomaterials such as 316 L stainless steel and cobalt–chromium alloys (higher than 200 GPa), it is still much higher than that of natural human bone (10–30 GPa) [4]. The modulus mismatch between implants and surrounding human bones can lead to a stress shielding effect, resulting in bone resorption and premature failure of the implant [5]. Additionally, the release of toxic Al and V ions from Ti-6Al-4V is associated with long-term health problems, such as Alzheimer disease, neuropathy and ostemomalacia [6]. Consequently, this has led to the development of β-type Ti alloys that consist of non-toxic alloying elements and process lower modulus than that of α- and ($\alpha + \beta$)-type Ti alloys [7–9].

The β-type Ti alloys can exhibit a martensitic transformation from the body centered cubic (bcc) β phase (space group, Im-3m) to the orthorhombic α'' phase (space group, Cmcm) [10,11]. The martensitic transformation temperature decreases with the increase in the concentration of alloying elements, and the single β phase can be kept to room temperature upon quenching when the concentration exceeds a critical value [12,13]. It has been well recognized that the elastic modulus of β phase in Ti alloys is closely related their phase stability, with lower modulus corresponding to lower phase stability [14]. Therefore, the concentration of β stabilizers in most low modulus β-type Ti alloys was carefully designed to be as low as possible while being slightly higher than the critical concentration, in order to stabilize the single β phase against α'' martensitic transformation [15]. These alloys with low phase stability (i.e., low modulus) also have various deformation mechanisms, e.g., stress-induced martensitic transformation (SIMT), deformation twinning, dislocation slip, etc. [16–19]. The various deformation mechanisms enable the alloys to possess unique mechanical properties involving shape memory effect [20], superelasticity [21], high strain hardening rate [22], large recoverable strain [23], and nonlinear elastic-like behavior [24]. Among these properties, nonlinear elasticity has attracted considerable attention, since it exists in several multifunctional Ti alloys including the Gum Metal and Ti24Nb4Zr8Sn alloy [25,26]. Although several reversible deformation mechanisms such as lattice distortion [27], nanodisturbance [28], dislocation loops [29], and strain glass transition [30], were proposed, it was generally accepted that SIMT plays an important role in this kind of peculiar deformation behavior [31,32].

Due to the reversibility of SIMT after the release of the stress, in-situ experiments provide a very efficient method to explore the deformation mechanism of metastable β-type Ti alloys. Currently, in-situ conventional X-ray diffraction (XRD) has been employed to detect stress-induced α'' martensite in Ti13Nb4Mo and Ti26Nb alloys [33,34]. However, the volume fraction of stress-induced α'' martensite is usually very low in Ti alloys with nonlinear elasticity, which makes it difficult or even impossible to characterize the SIMT through conventional XRD. Furthermore, it is difficult to separate the main peaks of the β phase and α'' martensite, as the laboratory X-ray sources have relatively great wavelength and the K_{a1} and K_{a2} wavelengths coexist. Especially, the identification of β phase and α'' martensite from conventional XRD patterns will become even harder for alloys subjected to severe cold deformation because of the broadening of diffraction peaks. By contrast, synchrotron X-ray diffraction (SXRD) technique can trace the formation of a small volume fraction of phases during deformation due to the combination of short wavelength, good monochromaticity, high penetration, low absorption, and high resolution [35–37]. In-situ SXRD have been used to study the nonlinear deformation behavior of the Ti24Nb0.5O (at.%), Ti24Nb4Zr8Sn and Gum Metal, and it is demonstrated that SIMT, to a small extent, really exists during loading and contributes to the nonlinear elasticity [38–40].

Recently, our group developed several TiNb-based alloys with the β stabilizer concentration below the critical value [41,42]. These alloys consist of β and α'' phases in solution treated and quenched states, suggesting the intrinsic low β phase stability. Upon cold rolling plus subsequent short time annealing treatment, single β phase was nearly obtained due to the suppression of martensitic transformation. Furthermore, the precipitates formed during annealing treatment did not result in a detectible increase in β stabilizers in residual β matrix, and thus the β stabilizer concentration of β matrix after thermo-mechanical treatment is identical to that in solution treated state [43]. As a result, even lower elastic modulus was realized in these alloys, e.g., 46 GPa for the Ti36Nb5Zr alloy and 36 GPa for the Ti33Nb4Sn alloys [41,43]. Interestingly, the Ti36Nb5Zr alloy subjected to cold rolling plus annealing treatment exhibits a nonlinear deformation behavior [44], similar to that of Ti24Nb4Zr8Sn and Gum metal. Since the Ti36Nb5Zr alloys have lower β phase stability than the low modulus Ti alloys with chemical composition above the critical concentration, SIMT might occur during loading and contribute to its deformation behavior. However, α'' martensite has not been observed by conventional XRD and the underlying mechanism for the peculiar deformation behavior remains ambiguous.

In this paper, in situ SXRD experiments during uniaxial tensile loading were performed to explore the deformation mechanisms in the Ti36Nb5Zr alloy. The one-dimensional (1D) and two-dimensional (2D) SXRD patterns were obtained from in situ measurements to characterize the microstructural evolution of the alloy during in situ loading. Special attention was focused on the evolution of lattice strains and relative integrated diffraction peak intensities of both the β and α'' phases as functions of macroscopic applied strain. Our results indicated that the peculiar deformation behavior was closely related to various kinds of deformation mechanisms including elastic deformation, SIMT and plastic deformation, which were activated at different external strains.

2. Materials and Methods

An ingot with a nominal composition of Ti36Nb5Zr (wt.%) was fabricated by arc melting in an argon atmosphere using high purity Ti (99.99%), Nb (99.95%) and Zr (99.95%). The ingot was re-melted four times in the furnace to obtain chemical composition homogeneity. The as-cast ingot was hot forged to a billet with a thickness of 8 mm and width of 60 mm, and then homogenized at 1223 K for 5 h in vacuum, followed by water quenching. The homogenized billet was cold rolled into a plate of approximately 1 mm in thickness with a final reduction ratio of 87.5%. The tensile specimens with the rolling direction parallel to the loading axis were cut from the cold rolled plate using an electro-discharge machine. These tensile specimens were annealed at 698 K for 20 min and finally quenched into water. Uniaxial tensile tests were conducted at a strain rate of 1×10^4 s^{-1} on an Instron 5982 machine using specimens with a gage length of 30 mm and a cross section of 1×1.46 mm^2. In order to ensure accuracy of strain, a strain extensometer was used to record the stress–strain curves.

In situ SXRD experiments were conducted during tensile loading on the 11-ID-C beamline at the Advanced Photon Source, Argonne National Laboratory. High-energy X-rays with an energy of 115 keV, wavelength of 0.10798 Å and beam size of 0.4×0.4 mm^2 were used in transmission geometry, as shown in the schematic set-up in Figure 1. A PerkinElmer large area detector (PerkinElmer Inc., Waltham, MA, USA) of 2048 × 2048 pixels with a spatial resolution of 200 μm (pixel size) was placed behind the sample to collect the 2D diffraction patterns. The loading direction (LD) and the beamline are parallel to the rolling direction (RD) and the normal direction (ND) of the rolled plate, respectively. The azimuth angles (φ: 0–360°) at the Debye rings are defined to be 0° and 90° at the transverse direction (TD) and the longitudinal direction (parallel to LD), respectively. Fit2d software was employed to process the 2D diffraction images, and standard CeO_2 powder was used for calibration. The 1D SXRD spectrums were obtained by integrating along specific azimuth angles over a range of ±10° in the 2D diffraction patterns. The positions and areas of 1D diffraction peaks were determined by Gaussian fit. The evolution of the interplanar spacing (d-spacing) with respect to the initial state is indicated by the lattice strain, i.e., $\varepsilon_{hkl} = (d_{hkl} - d^0_{hkl})/d^0_{hkl}$, where d_{hkl} is the interplanar spacing of the (hkl) crystal plane with an external stress. The d^0_{hkl} is determined from the d-spacing of stress-free sample for the β phase, and from the d-spacing of sample subject to the stress that is high enough to resolve the accurate position of the martensite peak for the α'' phase. The relative intensity is defined as the ratio of the integrated area of a peak to that at the strain-free state for the β phase and to that at the maximum applied strain for the α'' phase, respectively.

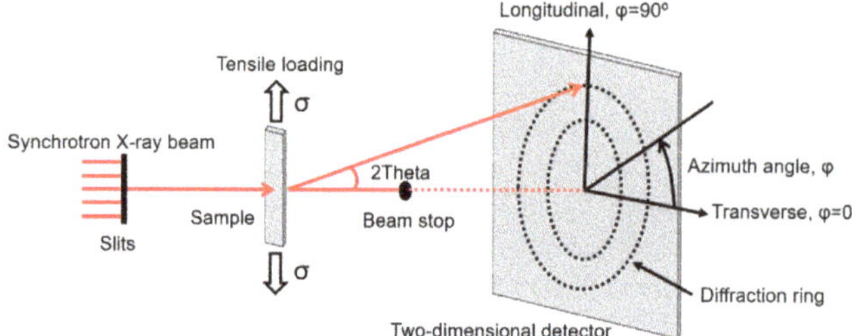

Figure 1. Schematic set-up of in situ synchrotron X-ray experiments.

3. Results

3.1. Microstructure and Macroscopic Mechanical Behavior

The microstructural evolution of Ti36Nb5Zr alloy during thermo-mechanical treatment has been described in detail in our previous work [41,44]. In brief, the Ti36Nb5Zr alloy after cold rolling and short time annealing treatment consists of a dominant β phase and a trace of a nanosized α phase. The annealing treatment did not result in significant recrystallization due to the low annealing temperature and short duration. The existence of high density of dislocations and grain boundaries suppressed the formation of α'' martensite in thermo-mechanically treated alloys, although the solution treated alloy consisted of dual ($\beta + \alpha''$) phases. Figure 2a,b present the 2D SXRD pattern and 1D SXRD spectrum obtained by integrating over the entire 360° of the Ti36Nb5Zr alloy before tensile testing. It can be seen that the intensity of peaks for the α phase is very weak in comparison to that for the β phase, verifying that the volume fraction of α phase is very low and thus its precipitation should not result in obvious chemical stabilization of the residual β matrix.

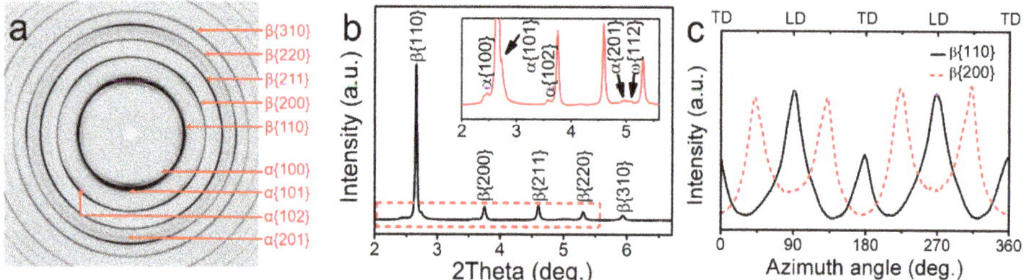

Figure 2. Structural analysis of the Ti36Nb5Zr alloy before tensile loading. (**a**) Two-dimensional SXRD pattern of the alloy. (**b**) One-dimensional SXRD spectrum integrated over the entire 360°. Inset shows the enlarged view of the boxed area in the spectrum. (**c**) Intensity distributions of the $\{110\}_\beta$ and $\{200\}_\beta$ diffraction peaks along the azimuth angle.

The uneven intensity distribution of the 2D SXRD pattern along different azimuth angles indicates that the alloy has a clear preferential orientation. The intensity distributions of the $\{110\}_\beta$ and $\{200\}_\beta$ diffraction peaks were plotted against the azimuth angle, as shown in Figure 2c. The maximum of the diffraction intensity for the $\{110\}_\beta$ peak appears at φ values of 0°, 90°, 180° and 360°, suggesting the existence of α-fiber texture components (i.e., grains with $<110>_\beta$ crystal direction parallel to RD). The azimuth angle of maximum diffraction intensity between the $\{110\}_\beta$ and $\{200\}_\beta$ peaks can be determined to be 46° ± 2°. Combined with the fact that the angle between the $\{110\}_\beta$ and $\{200\}_\beta$ crystal planes for bcc structure is 45°, it can be demonstrated that the texture component of the cold rolled and annealed Ti36Nb5Zr alloy is {001}<110> [36]. This kind of texture is commonly observed in β-type TiNb-based alloys subjected to cold rolling/annealing or warm rolling/annealing treatment, and is closely related to the martensitic transformation behavior [38,45].

Figure 3a shows the cyclic tensile stress–strain curves at an interval of 0.5% to a total strain of 3.5%. The loading and unloading curves to a strain of 1% are overlapped, and the 1.5% loading strain is almost fully recovered during unloading with a residual strain of only 0.03%. The recoverable strain increases with increasing applied external strain, e.g., 2.01% and 2.11% are achieved at a loading strain of 2.5% and 3.5%, respectively. It is worth noting that a nonlinear deformation behavior is clearly observed when the loading strain exceeds the linear elastic range limit of ~0.6%. The peculiar nonlinear deformation behavior as well as large recoverable strain might be attributed to the low β phase stability of the Ti36Nb5Zr alloy, since such a phenomenon is usually observed in metastable β-type Ti alloys [25,46]. Figure 3b present the tensile stress–strain curve during in situ SXRD experiment, and 2D diffraction patterns were taken at each block on the curve. The nonlinearity of the stress–strain curve upon in situ tensile loading is similar to that upon cyclic loading in Figure 3a. Furthermore, the in situ stress–strain curve can be divided into several stages by points O to D. OA (<0.67% strain) is undoubtedly the initial linear elastic deformation, while the mechanism of other stages will be discussed later. It is worth noting that no strain extensometer was used during the in-situ experiment, leading to overestimation of strains in Figure 3b. Therefore, the linear elastic range limit (~0.67%) in Figure 3b is slightly higher than that in Figure 3a (~0.6%).

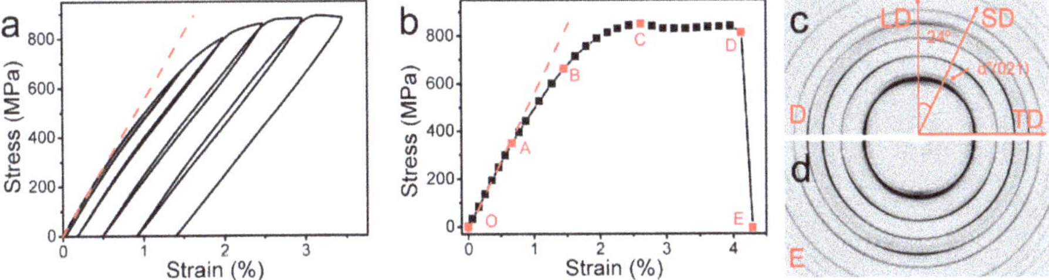

Figure 3. Mechanical behavior and SXRD patterns of the Ti36Nb5Zr alloy during tensile loading. (**a**) Cyclic stress–strain curves with 0.5% strain step. (**b**) Uniaxial tensile stress–strain curve during in situ SXRD experiment. The 2D SXRD patterns at different applied strains as noted in (**b**): (**c**) 4.13% (point D), (**d**) after fracture (point E). LD, TD and SD in (**c**,**d**) are abbreviations of loading direction, transverse direction and specific direction, respectively.

The 2D SXRD patterns of points D (corresponding to the maximum of external strain of 4.13%) and E (corresponding to the sample after fracture, i.e., the release of external strain) are shown in Figure 3c,d. Diffraction spots ascribed to the (021) crystal plane of α'' martensite can be clearly observed in diffraction rings at an applied strain of 4.13%, demonstrating the existence of SIMT. Moreover, the angle between the $(021)_{\alpha''}$ diffraction spots and the loading direction is about 24°, and this will be explained by the preferred selection of martensitic variants during SIMT in the next section. Therefore, the specific direction (SD) with an azimuth angle of 66° (i.e., 24° from loading direction) as well as the loading direction (LD, azimuth angle: 90°) will be selected to clarify the microscopic mechanisms of deformation for the present Ti36Nb5Zr alloy. After the release of applied strain, the α'' martensite disappeared and the 2D SXRD pattern was almost same with that before the tensile test (Figure 2a), indicating the complete reversibility of SIMT.

3.2. In Situ SXRD Characterization along the LD and SD

Figure 4a,b show the 1D SXRD spectrums during in situ tensile loading obtained by integrating along the longitudinal direction (φ: 80–100°) of the 2D SXRD patterns. Upon loading, the diffraction peaks of the β phase shift slightly towards lower Bragg angles, demonstrating a tensile elastic deformation in the LD. The $(021)_{\alpha''}$ and $(222)_{\alpha''}$ diffraction peaks are present at an applied strain of 0.88% and 1.28%, respectively. With the increase in external strain, the diffraction peaks of α'' martensite intensified, indicating a progressive transformation.

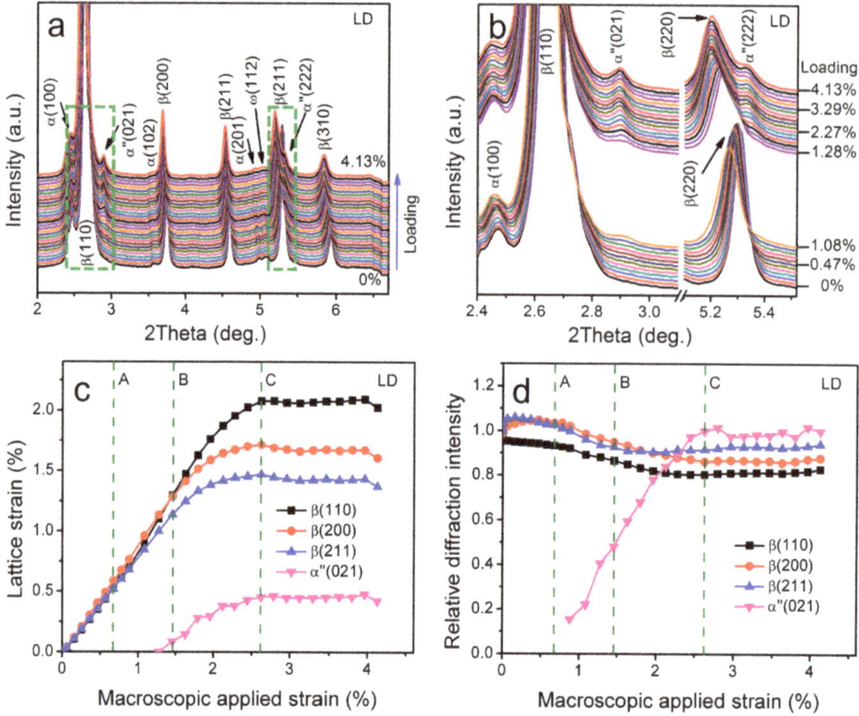

Figure 4. In situ SXRD analysis of the microstructural characteristics of the Ti36Nb5Zr alloy along LD. (**a**) One-dimensional SXRD spectrums. (**b**) Enlarged views of the boxed areas in (**a**). Evolution of the (**c**) lattice strains and (**d**) relative integrated diffraction peak intensities of the $(110)_\beta$, $(200)_\beta$, $(211)_\beta$ and $(021)_{\alpha''}$ crystal planes as functions of the macroscopic applied strain. A, B and C in (**c**,**d**) represent the different macroscopic strains noted in Figure 3b.

Figure 4c,d show the evolution of the lattice strains and relative integrated diffraction peak intensities in the LD for the $(110)_\beta$, $(200)_\beta$, $(211)_\beta$ and $(021)_{\alpha''}$ peaks during in situ tensile loading. The d^0_{021} of α'' martensite is defined to be the *d*-spacing at an applied strain of 1.28%, since the $(021)_{\alpha''}$ peaks at lower external strains are too weak to be fitted for accurate peak positions and it is also difficult to obtain the *d*-spacing of martensite under zero external stress due to the complete reversibility of SIMT in the present alloy. In the stage of O–A–B (applied strain range: 0–1.46%), the lattice strains of all β crystal planes increase linearly with the increase in external strain, implying a elastic deformation behavior; in the stage of B–C (applied strain range: 1.46–2.62%), the lattice strains of the β crystal planes continue to increase at a much reduced rate with further increase in the applied strain, indicating the commencement of plastic deformation; in the stage of C–D (applied strain range: 2.62–4.13%), the lattice strains of both the β and α'' crystal planes remain almost constant, suggesting a complete stop of elastic deformation in the local area under SXRD study. The relative diffraction intensity reveals that progressive SIMT occurred with an external strain of up to 2.62% (point C), which can be demonstrated by the continuous increase in relative intensity of $(021)_{\alpha''}$ at the expense of that of $(110)_\beta$, $(200)_\beta$ and $(211)_\beta$. It is worth noting that the nonlinear deformation behavior appears at a strain of 0.67% (point A) while the SIMT along LD is first observed at a strain of 0.88%. The martensitic variants characterized by the $(021)_{\alpha''}$ peak along the LD might not form firstly during tensile loading; the examination of martensitic transformation along other azimuth angles should be considered.

Figure 5a,b show the 1D SXRD spectrums during in situ tensile loading obtained by integrating along a specific direction (SD, φ: 56–76°) of the 2D SXRD patterns. The shift of the diffraction peaks of the β phase towards lower angles indicates that a tensile elastic deformation exits in the SD. The diffraction peaks ascribed to $(021)_{\alpha''}$ and $(222)_{\alpha''}$ started to appear at certain external strain values, and intensified with the increase in applied strain, demonstrating that gradual SIMT occurred during tensile loading. However, the applied strains for the first appearance of $(021)_{\alpha''}$ and $(222)_{\alpha''}$ diffraction peaks are 0.67% and 1.46%, respectively, which is different from the results of 1D XRD spectrums along the LD. This implies that the β phase grains with different crystal orientations have different critical stress for SIMT. Besides, the intensity of α'' martensite is clearly greater in the SD than that in the LD, suggesting the preferred selection of martensite variants.

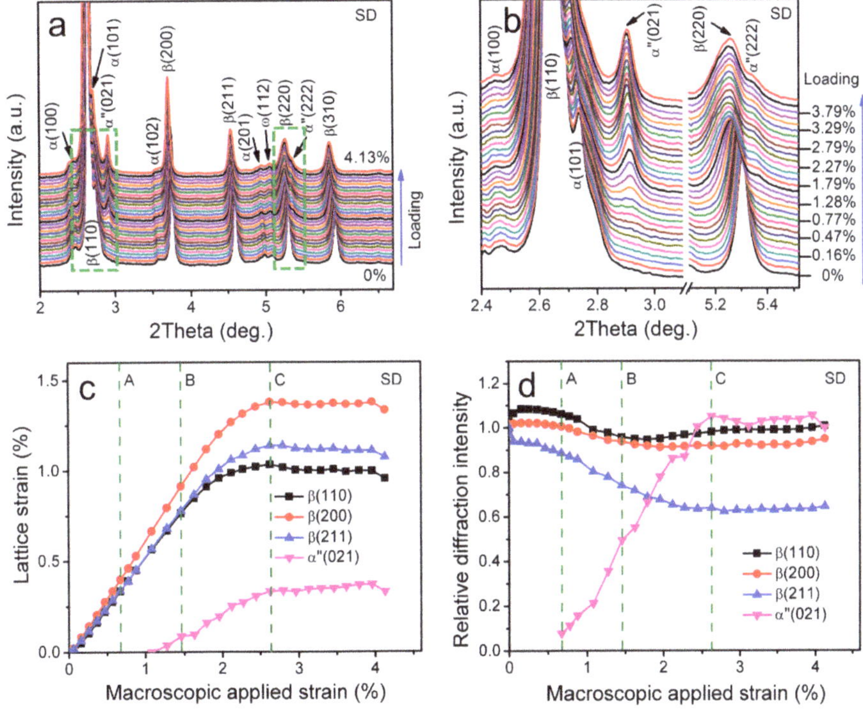

Figure 5. In situ SXRD analysis of the microstructural characteristics of the Ti36Nb5Zr alloy along SD. (**a**) One-dimensional SXRD spectrums. (**b**) Enlarged views of the boxed areas in (**a**). Evolution of the (**c**) lattice strains and (**d**) relative integrated diffraction peak intensities of the $(110)_\beta$, $(200)_\beta$, $(211)_\beta$ and $(021)_{\alpha''}$ crystal planes as functions of the macroscopic applied strain. A, B and C in (**c**,**d**) represent different macroscopic strains noted in Figure 3b.

Figure 5 show the evolution of the lattice strains and relative integrated diffraction peak intensities for the $(110)_\beta$, $(200)_\beta$, $(211)_\beta$ and $(021)_{\alpha''}$ as functions of applied strain in the SD. The d^0_{021} of α'' martensite is defined to be the d-spacing at an applied strain of 0.88% when the $(021)_{\alpha''}$ diffraction peak is strong enough to be fitted for accurate peak position. The evolution of the lattice strains in the SD is similar to that in the LD. In brief, in the stage of O-A-B, the lattice strain–macroscopic applied strain curves are linear for the β phase, reflecting an elastic deformation; in the stage of B-C, the lattice strain–macroscopic applied strain curves deviate from the linearity and the slopes of the curves begin to decrease, indicating a elastoplastic deformation; in the stage of C-D, the lattice strains of all crystal planes remain almost unchanged, suggesting that elastic deformation disappears in the local area under SXRD study. The evolution of the relative integrated intensity of the $(021)_{\alpha''}$ diffraction peak indicates that the onset of SIMT corresponds to point A in the tensile stress–strain curve in Figure 3b, which means the deviation from linearity observed in macroscopic stress–strain curve is coincident with the SIMT. This provides direct evidence that the nonlinear deformation behavior of the cold rolled and annealed Ti36Nb5Zr alloy could be attributed to the SIMT during loading.

4. Discussion

4.1. Variant Selection of Stress-Induced Martensite

To further explore the detailed scenarios of SIMT, the 2D SXRD patterns at different applied strains (corresponding to points O–E in the macroscopic stress–strain curve in Figure 3b) were unrolled along the full azimuthal circle (0°–360°) and presented in Figure 6. The diffraction lines of the β phase are non-uniform and even discontinuous at zero external strain (O), indicating the existence of strong texture as described above. The diffraction lines curved into "banana" shapes with increasing the applied strain (A–D), indicating that the specimen experiences maximum tension and compression in the longitudinal direction (φ: 90° and 270°) and the transverse direction (φ: 0° and 180°), respectively. Faint shadows ascribed to $(021)_{\alpha''}$ appeared at the applied strain corresponding to point A, and evolved into diffraction spots at specific azimuth angles with the increase in external strain (B–D), implying the progressive SIMT during loading. By contrast, the diffraction spots of $(222)_{\alpha''}$ formed at higher applied strain and are weaker than those of $(021)_{\alpha''}$. In addition to the diffraction spots of $(021)_{\alpha''}$ and $(222)_{\alpha''}$, no α'' diffraction spots can be identified from the unrolled 2D SXRD images. Furthermore, it should be emphasized that the intensity of α'' martensite is much lower than that of the parent β phase even at the maximum applied strain (point D), which is evidenced by the 1D SXRD spectrums integrated over the entire 360° shown in Figure 7. This suggests the transformed fraction of the β phase is very low, which might be the reason why a nonlinear deformation instead of a yielding plateau is observed during loading of the present Ti36Nb5Zr alloy.

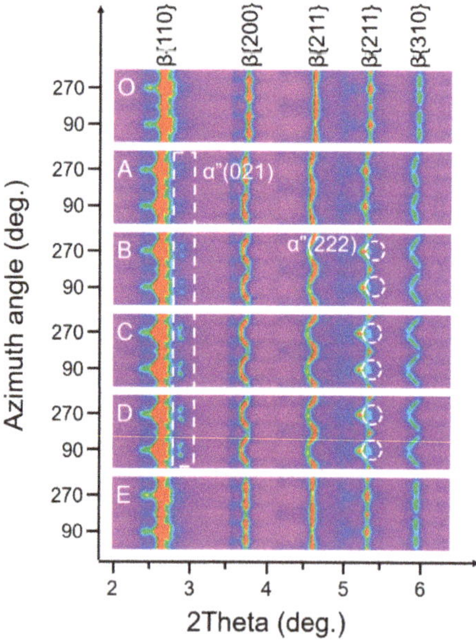

Figure 6. The unrolled 2D SXRD images along the azimuthal circle (0° to 360°) at different applied strains corresponding to points O–E in Figure 3b.

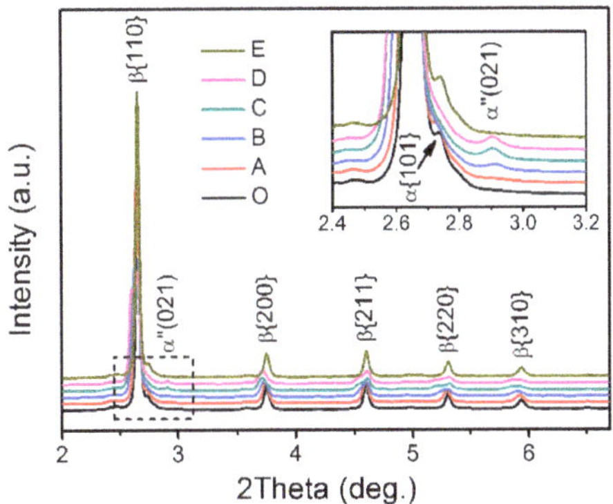

Figure 7. One-dimensional SXRD spectrums integrated over the entire 360° at different applied strains corresponding to points O–E in Figure 3b. Inset shows the enlarged view of the boxed area in the spectrums.

Figure 8 shows the intensity distribution of the $(021)_{\alpha''}$ diffraction peak along the azimuth angle at different applied strains corresponding to points O–E in Figure 3b. The curves at applied strains corresponding to points O and E are overlapped, demonstrating the complete reversibility of SIMT in the present alloy. Moreover, the overlap of curves corresponding to points C and D reveals that SIMT did not further occur when the external strain exceeded point C in the local area under SXRD study. The angle between $(021)_{\alpha''}$ diffraction peaks and the LD is determined to be 24° ± 3° at the maximum applied strain (corresponding to point D). From the energetics of $\beta \rightarrow \alpha''$ transformation, the critical stress of SIMT depends on the initial orientation of β grains, and the minimum critical stress can be realized in the β grains that are orientated with a <110> direction along the tensile direction [40]. Moreover, it has been reported that a variant selection operates during SIMT process and only the variants that give a maximum of transformation strain can be formed [36,47]. As mentioned above, the present Ti36Nb5Zr alloy has a texture <110> fiber texture component, i.e., $<110>_\beta$ is parallel to the rolling direction/LD. Therefore, SIMT will first occur in the β grains with <110> parallel to the tensile axis due to their having the lowest critical stress. Furthermore, only one martensitic variant that gives the maximum transformation is activated. The $[100]_{\alpha''}$, $[10]_{\alpha''}$ and $[1]_{\alpha''}$ crystal orientation of this specific variant are parallel to $[1]_\beta$, $[110]_\beta$ and $[1\text{–}10]_\beta$, respectively. This implies that the $(110)_\beta$ and $(020)_{\alpha''}$ peaks will appear in the LD of the 2D SXRD patterns and the positions of other α'' peaks can be calculated according to the lattice parameters. As only two α'' peaks were observed in our measurements, the lattice parameters of α'' martensite cannot be calculated for the present alloy. As a solution, the lattice parameters of Gum Metal, which has a similar Nb and Zr content with the Ti36Nb5Zr, were used here, i.e., a = 3.250 Å, b = 4.853 Å and c = 4.740 Å [32]. Based on this assumption, the angle between $[20]_{\alpha''}$ and $[21]_{\alpha''}$ is calculated to be 26°, i.e., the angle between the $(021)_{\alpha''}$ peaks and the LD in 2D SXRD diffraction patterns is 26°, which is consistent with the experimental value (24° ± 3°).

As mentioned before, the present Ti36Nb5Zr alloy did not experience complete recrystallization during the annealing process, thus the existence of large amount of defects such as dislocations and grain boundaries resulted in the broadening of β diffraction peaks. Furthermore, the volume fraction of α'' martensite is much lower than that of the β phase. Consequently, most α'' peaks were either overlapped with β peaks or too weak to be identified from diffraction patterns. According to the PDF card (No. 17-0102), the $(021)_{\alpha''}$ peak is one of the second strongest peaks of the α'' phase, and the distance between the $(021)_{\alpha''}$ peak and the β diffraction peaks is relatively large. This might be the reason why only $(021)_{\alpha''}$ diffraction peaks of martensite transformed from β grains with <110> parallel to the tensile axis were observed. In the case of $(021)_{\alpha''}$ peaks along the LD and $(222)_{\alpha''}$ peaks along both the LD and SD, these martensite variants were transformed from β grains whose $<110>_\beta$ direction is not parallel to the tensile axis. Therefore, higher critical stresses for SIMT are required, which agrees well with the experimental results that the first appearance of these martensitic peaks occurred at higher external strain than the $(021)_{\alpha''}$ peak along the LD. Furthermore, the intensities of these peaks are relatively weaker due to the α-fiber texture component in the original β grains.

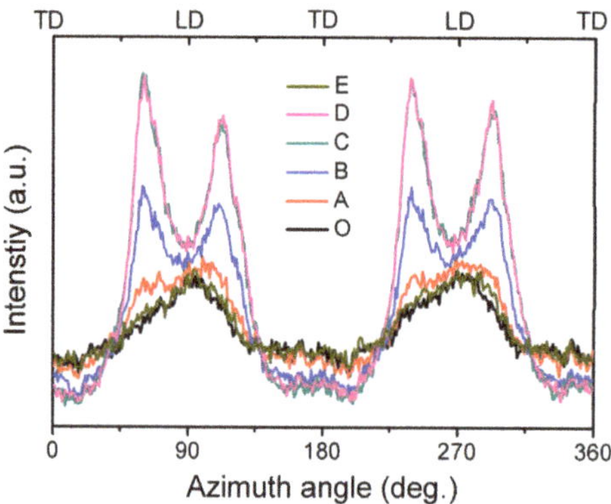

Figure 8. Intensity distributions of the $(021)_{\alpha''}$ diffraction peaks along the azimuth angle at different applied strains corresponding to points O–E in Figure 3b.

4.2. Origin of the Recoverable Strain

The cyclic tensile stress–strain curve in Figure 3a indicates that the cold rolled and annealed Ti36Nb5Zr alloy processes a maximum recoverable strain of 2.11%. This is much larger than that of most engineering materials (<0.5%) and is even similar to that of bulk metallic glasses [48–51], although it is lower than that of superelastic β-Ti alloys whose recoverable strain is mainly realized by SIMT [11]. Considering that the volume fraction of the transformed α'' martensite is very low, the direct contribution of SIMT to the recoverable strain in the present alloy should be small. Figure 4c indicates that the lattice strains along the LD for the $(110)_\beta$, $(200)_\beta$ and $(211)_\beta$ reached maximum values of 2.08%, 1.71% and 1.41% at applied strain of 2.62% (corresponding to point C in macroscopic stress–strain in Figure 3b). It is worth noting that the maximum lattice strain of the $(110)_\beta$ at point C is close to the macroscopic recoverable strain (2.01%) at a similar applied strain (2.5%) to those determined from the cyclic stress–strain curves in Figure 3a. Considering that the present alloy exhibits a strong <110> fiber texture (i.e., $(110)_\beta$ perpendicular to the LD), it is proposed that the recoverable strain of the present Ti36Nb5Zr alloy is mainly contributed by the elastic strain of the β phase. It has been reported that the martensitic transforming alloy can exhibit much larger elastic strain than the conventional dislocation slip alloy [52]. The possibility of SIMT implies the structural instability of the parent phase, and the uniform lattice distortion provided by martensitic transformation can suppress strain localization and damage accumulation [53]. These two characteristics enable alloys that can undergo SIMT to possess large elastic strain. The present Ti36Nb5Zr alloy was designed to have low β phase stability in order to realize low modulus, which provides the possibility of the occurrence of SIMT. In other words, the low β phase stability leads to the large elastic strain that dominated the large recoverable strain of the alloy.

4.3. Microscopic Deformation Mechanisms at Different Macroscopic Applied Strains

Based on the evolution of lattice strains and relative integrated diffraction peak intensities of both the β and α'' phases, it is possible to elucidate the activation sequence of each deformation mechanism at different applied strains. This sequence can be summarized on the cyclic and in situ tensile stress–strain curves of the Ti36Nb5Zr alloy, as shown in Figure 9. In the stage of O–A, the deformation was only accommodated by the elastic deformation of the β phase, which corresponds to linear elastic range in the stress–strain curves. In the stage of A–B, SIMT progressively occurred with increasing external strain and the onset of SIMT at point A corresponds to the start of nonlinearity in the stress–strain curves. Besides, the elastic deformation of the β phase continued during this stage. The elastic deformation as well as the reversible SIMT contributed to the fully recoverable strain of 1.5% in the cyclic tensile loading. In the stage of B–C, the SIMT process continued while the β exhibited elastic and plastic deformation simultaneously. These mechanisms provided a ~2.01% recoverable strain at an applied strain of 2.5% during cyclic tensile loading.

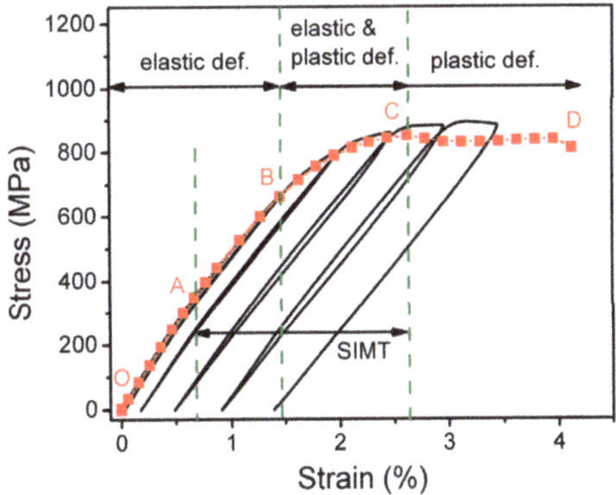

Figure 9. Domains of occurrence of different deformation mechanisms noted on conventional stress–strain curves for Ti36Nb5Zr alloy. "def" is the abbreviation of "deformation".

In the stage of C–D, the homogeneous plastic deformation evolved into inhomogeneous plastic deformation due to the lack of strain hardening, resulting in plastic strain localization. Romanova et al. have reported that the local strain in the subsection that is far from the one where a neck will form, ceases to develop as soon as plastic strain localizes [54]. Actually, the tensile specimens break near one of the movable grips rather than the center where the synchrotron X-ray beam penetrated into the sample. Therefore, it is believed the plastic deformation of the local area under SXRD study did not continue when the applied strain exceeded point C. That might be why the lattice strains and the relative integrated intensities of all crystal planes remain almost unchanged in the stage of C–D. Although the SXRD experiment was not carried out in the local area of neck formation, it is believed that the plastic deformation continued in this region. On the other hand, the reversible deformation mechanisms including elastic deformation and/or SIMT existed but contributed little to deformation behavior due to the slight increase in the recoverable strain in the stage of C–D. Therefore, it is proposed that plastic deformation dominated the process in the stage of C–D.

5. Conclusions

The deformation mechanisms of the low modulus Ti36Nb5Zr alloy were investigated by in situ SXRD experiments. The following conclusions can be drawn:

(1) The cold rolled plus annealed Ti36Nb5Zr alloy consists of dominant β phase and a trace of α phase, and has a {001}<110> texture component. During loading, a nonlinear deformation behavior appeared after the loading strain exceeded the linear range limit of ~0.6%, and large recoverable strains of 2.01% and 2.11% were obtained at applied strains of 2.5% and 3.5%, respectively.

(2) SIMT occurred at an external strain of 0.67% and continued with an applied strain up to 2.67%. Furthermore, the onset of SIMT corresponds to the beginning of the nonlinearity in macroscopic stress–strain curves, indicating that the nonlinear deformation behavior originates from the SIMT. Besides, the α'' diffraction spots only appeared at specific azimuth angles on the 2D SXRD patterns, which was caused by the preferred orientation of the original β grains and the stress-induced selections of martensitic variants.

(3) The large recoverable strain was dominated by the elastic deformation of the β phase, which resulted from the low β phase stability of the Ti36Nb5Zr alloy, whereas the $\beta \rightarrow \alpha''$ phase transformation strain contributed little to the recoverable strain due to the low volume fraction of the transformed β phase.

(4) Various deformation mechanisms were activated at different applied strains, including elastic deformation at applied strains of 0–1.46%, SIMT at applied strains of 0.67–2.62%, elastoplastic deformation at applied strains of 1.46–2.62% and plastic deformation at applied strains exceeding 2.62%.

Author Contributions: Conceptualization, Q.M. and X.Z.; methodology, Q.M.; validation, J.Q. and F.W.; formal analysis, K.W.; investigation, H.L.; resources, Y.S. and B.S.; data curation, H.L.; writing—original draft preparation, H.L.; writing—review and editing, Q.M. and X.Z.; visualization, H.L.; supervision, X.Z.; project administration, S.G.; funding acquisition, Q.M. All authors have read and agreed to the published version of the manuscript.

Funding: This research was funded by the Fundamental Research Funds for the Central Universities, grant number 2017QNA04.

Acknowledgments: Qingkun Meng thanks Cun Yu for technical assistance in the in situ synchrotron experiments, and Junsong Zhang for helpful discussions on the deformation mechanisms. The use of the Advanced Photon Source was supported by the U.S. Department of Energy, Office of Science, and Office of Basic Energy Science under Contract No. DE-AC02-06CH11357.

Conflicts of Interest: The authors declare no conflict of interest.

References

1. Niinomi, M.; Nakai, M.; Hieda, J. Development of new metallic alloys for biomedical applications. *Acta Biomater.* **2012**, *8*, 3888–3903. [CrossRef]
2. Chen, L.-Y.; Cui, Y.-W.; Zhang, L.-C. Recent Development in Beta Titanium Alloys for Biomedical Applications. *Metals* **2020**, *10*, 1139. [CrossRef]
3. Niinomi, M. Mechanical biocompatibilities of titanium alloys for biomedical applications. *J. Mech. Behav. Biomed. Mater.* **2008**, *1*, 30–42. [CrossRef] [PubMed]
4. Zhang, L.-C.; Chen, L.-Y. A Review on Biomedical Titanium Alloys: Recent Progress and Prospect. *Adv. Eng. Mater.* **2019**, *21*, 1801215. [CrossRef]
5. Sumner, D.R.; Turner, T.M.; Igloria, R.; Urban, R.M.; Galante, J.O. Functional adaptation and ingrowth of bone vary as a function of hip implant stiffness. *J. Biomech.* **1998**, *31*, 909–917. [CrossRef]
6. Geetha, M.; Singh, A.K.; Asokamani, R.; Gogia, A.K. Ti based biomaterials, the ultimate choice for orthopaedic implants—A review. *Prog. Mater. Sci.* **2009**, *54*, 397–425. [CrossRef]
7. Li, P.; Ma, X.; Jia, Y.; Meng, F.; Tang, L.; He, Z. Microstructure and Mechanical Properties of Rapidly Solidified β-Type Ti–Fe–Sn–Mo Alloys with High Specific Strength and Low Elastic Modulus. *Metals* **2019**, *9*, 1135. [CrossRef]
8. Li, P.; Ma, X.; Wang, D.; Zhang, H. Microstructural and Mechanical Properties of β-Type Ti–Nb–Sn Biomedical Alloys with Low Elastic Modulus. *Metals* **2019**, *9*, 712. [CrossRef]
9. Preisler, D.; Janeček, M.; Harcuba, P.; Džugan, J.; Halmešová, K.; Málek, J.; Veverková, A.; Stráský, J. The Effect of Hot Working on the Mechanical Properties of High Strength Biomedical Ti-Nb-Ta-Zr-O Alloy. *Materials* **2019**, *12*, 4233. [CrossRef]
10. Biesiekierski, A.; Wang, J.; Abdel-Hady Gepreel, M.; Wen, C. A new look at biomedical Ti-based shape memory alloys. *Acta Biomater.* **2012**, *8*, 1661–1669. [CrossRef]
11. Ramezannejad, A.; Xu, W.; Xiao, W.L.; Fox, K.; Liang, D.; Qian, M. New insights into nickel-free superelastic titanium alloys for biomedical applications. *Curr. Opin. Solid State Mater.* **2019**, *23*, 100783. [CrossRef]
12. Kim, H.Y.; Ikehara, Y.; Kim, J.I.; Hosoda, H.; Miyazaki, S. Martensitic transformation, shape memory effect and superelasticity of Ti–Nb binary alloys. *Acta Mater.* **2006**, *54*, 2419–2429. [CrossRef]
13. Kolli, R.P.; Devaraj, A. A Review of Metastable Beta Titanium Alloys. *Metals* **2018**, *8*, 506. [CrossRef]
14. Abdel-Hady, M.; Hinoshita, K.; Morinaga, M. General approach to phase stability and elastic properties of β-type Ti-alloys using electronic parameters. *Scr. Mater.* **2006**, *55*, 477–480. [CrossRef]

15. Banerjee, D.; Williams, J.C. Perspectives on Titanium Science and Technology. *Acta Mater.* **2013**, *61*, 844–879. [CrossRef]
16. Liang, Q.; Kloenne, Z.; Zheng, Y.; Wang, D.; Antonov, S.; Gao, Y.; Hao, Y.; Yang, R.; Wang, Y.; Fraser, H.L. The role of nano-scaled structural non-uniformities on deformation twinning and stress-induced transformation in a cold rolled multifunctional β-titanium alloy. *Scr. Mater.* **2020**, *177*, 181–185. [CrossRef]
17. Sun, F.; Zhang, J.Y.; Marteleur, M.; Gloriant, T.; Vermaut, P.; Laillé, D.; Castany, P.; Curfs, C.; Jacques, P.J.; Prima, F. Investigation of early stage deformation mechanisms in a metastable β titanium alloy showing combined twinning-induced plasticity and transformation-induced plasticity effects. *Acta Mater.* **2013**, *61*, 6406–6417. [CrossRef]
18. Chen, Q.J.; Ma, S.Y.; Wang, S.Q. The Nucleation and the Intrinsic Microstructure Evolution of Martensite from {332}<113>(beta) Twin Boundary in beta Titanium: First-Principles Calculations. *Metals* **2019**, *9*, 1202. [CrossRef]
19. Hanada, S.; Izumi, O. Correlation of tensile properties, deformation modes, and phase stability in commercial β-phase titanium alloys. *Met. Mater. Trans. A* **1987**, *18*, 265–271. [CrossRef]
20. Li, S.; Choi, M.-s.; Nam, T.-h. Effect of thermo-mechanical treatment on microstructural evolution and mechanical properties of a superelastic Ti–Zr-based shape memory alloy. *Mater. Sci. Eng. A* **2020**, *789*, 139664. [CrossRef]
21. Miyazaki, S.; Kim, H.Y.; Hosoda, H. Development and characterization of Ni-free Ti-base shape memory and superelastic alloys. *Mater. Sci. Eng. A* **2006**, *438–440*, 18–24. [CrossRef]
22. Zhang, J.; Sun, F.; Chen, Z.; Yang, Y.; Shen, B.; Li, J.; Prima, F. Strong and ductile beta Ti–18Zr–13Mo alloy with multimodal twinning. *Mater. Res. Lett.* **2019**, *7*, 251–257. [CrossRef]
23. Fu, J.; Yamamoto, A.; Kim, H.Y.; Hosoda, H.; Miyazaki, S. Novel Ti-base superelastic alloys with large recovery strain and excellent biocompatibility. *Acta Biomater.* **2015**, *17*, 56–67. [CrossRef]
24. Hao, Y.L.; Li, S.J.; Sun, S.Y.; Zheng, C.Y.; Hu, Q.M.; Yang, R. Super-elastic titanium alloy with unstable plastic deformation. *Appl. Phys. Lett.* **2005**, *87*, 091906. [CrossRef]
25. Saito, T.; Furuta, T.; Hwang, J.-H.; Kuramoto, S.; Nishino, K.; Suzuki, N.; Chen, R.; Yamada, A.; Ito, K.; Seno, Y.; et al. Multifunctional alloys obtained via a dislocation-free plastic deformation mechanism. *Science* **2003**, *300*, 464–467. [CrossRef]
26. Hao, Y.L.; Li, S.J.; Sun, S.Y.; Zheng, C.Y.; Yang, R. Elastic deformation behaviour of Ti–24Nb–4Zr–7.9Sn for biomedical applications. *Acta Biomater.* **2007**, *3*, 277–286. [CrossRef]
27. Zhang, Y.W.; Li, S.J.; Obbard, E.G.; Wang, H.; Wang, S.C.; Hao, Y.L.; Yang, R. Elastic properties of Ti–24Nb–4Zr–8Sn single crystals with bcc crystal structure. *Acta Mater.* **2011**, *59*, 3081–3090. [CrossRef]
28. Gutkin, M.Y.; Ishizaki, T.; Kuramoto, S.; Ovid'ko, I.A. Nanodisturbances in deformed Gum Metal. *Acta Mater.* **2006**, *54*, 2489–2499. [CrossRef]
29. Cui, J.P.; Hao, Y.L.; Li, S.J.; Sui, M.L.; Li, D.X.; Yang, R. Reversible Movement of Homogenously Nucleated Dislocations in a β-Titanium Alloy. *Phys. Rev. Lett.* **2009**, *102*, 045503. [CrossRef]
30. Liang, Q.; Wang, D.; Zheng, Y.; Zhao, S.; Gao, Y.; Hao, Y.; Yang, R.; Banerjee, D.; Fraser, H.L.; Wang, Y. Shuffle-nanodomain regulated strain glass transition in Ti-24Nb-4Zr-8Sn alloy. *Acta Mater.* **2020**, *186*, 415–424. [CrossRef]
31. Liu, J.P.; Wang, Y.D.; Hao, Y.L.; Wang, H.L.; Wang, Y.; Nie, Z.H.; Su, R.; Wang, D.; Ren, Y.; Lu, Z.P.; et al. High-energy X-ray diffuse scattering studies on deformation-induced spatially confined martensitic transformations in multifunctional Ti–24Nb–4Zr–8Sn alloy. *Acta Mater.* **2014**, *81*, 476–486. [CrossRef]
32. Talling, R.J.; Dashwood, R.J.; Jackson, M.; Dye, D. On the mechanism of superelasticity in Gum metal. *Acta Mater.* **2009**, *57*, 1188–1198. [CrossRef]
33. Al-Zain, Y.; Kim, H.Y.; Koyano, T.; Hosoda, H.; Nam, T.H.; Miyazaki, S. Anomalous temperature dependence of the superelastic behavior of Ti–Nb–Mo alloys. *Acta Mater.* **2011**, *59*, 1464–1473. [CrossRef]
34. Tahara, M.; Kim, H.Y.; Hosoda, H.; Miyazaki, S. Cyclic deformation behavior of a Ti–26 at.% Nb alloy. *Acta Mater.* **2009**, *57*, 2461–2469. [CrossRef]
35. Zhao, C.H.; Kisslinger, K.; Huang, X.J.; Lu, M.; Camino, F.; Lin, C.H.; Yan, H.F.; Nazaretski, E.; Chu, Y.; Ravel, B.; et al. Bi-continuous pattern formation in thin films via solid-state interfacial dealloying studied by multimodal characterization. *Mater. Horiz.* **2019**, *6*, 1991–2002. [CrossRef]

36. Zhu, Z.W.; Xiong, C.Y.; Wang, J.; Li, R.G.; Ren, Y.; Wang, Y.D.; Li, Y. In situ synchrotron X-ray diffraction investigations of the physical mechanism of ultra-low strain hardening in Ti-30Zr-10Nb alloy. *Acta Mater.* **2018**, *154*, 45–55. [CrossRef]
37. Warchomicka, F.; Canelo-Yubero, D.; Zehetner, E.; Requena, G.; Stark, A.; Poletti, C. In-Situ Synchrotron X-Ray Diffraction of Ti-6Al-4V During Thermomechanical Treatment in the Beta Field. *Metals* **2019**, *9*, 862. [CrossRef]
38. Castany, P.; Ramarolahy, A.; Prima, F.; Laheurte, P.; Curfs, C.; Gloriant, T. In situ synchrotron X-ray diffraction study of the martensitic transformation in superelastic Ti-24Nb-0.5N and Ti-24Nb-0.5O alloys. *Acta Mater.* **2015**, *88*, 102–111. [CrossRef]
39. Yang, Y.; Castany, P.; Cornen, M.; Prima, F.; Li, S.J.; Hao, Y.L.; Gloriant, T. Characterization of the martensitic transformation in the superelastic Ti–24Nb–4Zr–8Sn alloy by in situ synchrotron X-ray diffraction and dynamic mechanical analysis. *Acta Mater.* **2015**, *88*, 25–33. [CrossRef]
40. Morris, J.W.; Hanlumyuang, Y.; Sherburne, M.; Withey, E.; Chrzan, D.C.; Kuramoto, S.; Hayashi, Y.; Hara, M. Anomalous transformation-induced deformation in ⟨110⟩ textured Gum Metal. *Acta Mater.* **2010**, *58*, 3271–3280. [CrossRef]
41. Meng, Q.; Guo, S.; Ren, X.; Xu, H.; Zhao, X. Possible contribution of low shear modulus C44 to the low Young's modulus of Ti-36Nb-5Zr alloy. *Appl. Phys. Lett.* **2014**, *105*, 131907. [CrossRef]
42. Meng, Q.; Zhang, J.; Huo, Y.; Sui, Y.; Zhang, J.; Guo, S.; Zhao, X. Design of low modulus β-type titanium alloys by tuning shear modulus C44. *J. Alloys Compd.* **2018**, *745*, 579–585. [CrossRef]
43. Guo, S.; Meng, Q.; Zhao, X.; Wei, Q.; Xu, H. Design and fabrication of a metastable β-type titanium alloy with ultralow elastic modulus and high strength. *Sci. Rep.* **2015**, *5*, 14688. [CrossRef]
44. Meng, Q.; Liu, Q.; Guo, S.; Zhu, Y.; Zhao, X. Effect of thermo-mechanical treatment on mechanical and elastic properties of Ti–36Nb–5Zr alloy. *Prog. Nat. Sci.* **2015**, *25*, 229–235. [CrossRef]
45. Sander, B.; Raabe, D. Texture inhomogeneity in a Ti–Nb-based β-titanium alloy after warm rolling and recrystallization. *Mater. Sci. Eng. A* **2008**, *479*, 236–247. [CrossRef]
46. Hou, F.Q.; Li, S.J.; Hao, Y.L.; Yang, R. Nonlinear elastic deformation behaviour of Ti–30Nb–12Zr alloys. *Scr. Mater.* **2010**, *63*, 54–57. [CrossRef]
47. Tahara, M.; Kim, H.Y.; Inamura, T.; Hosoda, H.; Miyazaki, S. Lattice modulation and superelasticity in oxygen-added β-Ti alloys. *Acta Mater.* **2011**, *59*, 6208–6218. [CrossRef]
48. Sun, B.B.; Sui, M.L.; Wang, Y.M.; He, G.; Eckert, J.; Ma, E. Ultrafine composite microstructure in a bulk Ti alloy for high strength, strain hardening and tensile ductility. *Acta Mater.* **2006**, *54*, 1349–1357. [CrossRef]
49. Chen, X.H.; Lu, J.; Lu, L.; Lu, K. Tensile properties of a nanocrystalline 316L austenitic stainless steel. *Scr. Mater.* **2005**, *52*, 1039–1044. [CrossRef]
50. Jeon, C.; Kim, C.P.; Joo, S.H.; Kim, H.S.; Lee, S. High tensile ductility of Ti-based amorphous matrix composites modified from conventional Ti–6Al–4V titanium alloy. *Acta Mater.* **2013**, *61*, 3012–3026. [CrossRef]
51. Zhang, J.; Liu, Y.; Yang, H.; Ren, Y.; Cui, L.; Jiang, D.; Wu, Z.; Ma, Z.; Guo, F.; Bakhtiari, S.; et al. Achieving 5.9% elastic strain in kilograms of metallic glasses: Nanoscopic strain engineering goes macro. *Mater. Today* **2020**, *37*, 18–26. [CrossRef]
52. Zhang, J.; Hao, S.; Jiang, D.; Huan, Y.; Cui, L.; Liu, Y.; Ren, Y.; Yang, H. Dual Phase Synergy Enabled Large Elastic Strains of Nanoinclusions in a Dislocation Slip Matrix Composite. *Nano Lett.* **2018**, *18*, 2976–2983. [CrossRef]
53. Zhang, J.; Liu, Y.; Cui, L.; Hao, S.; Jiang, D.; Yu, K.; Mao, S.; Ren, Y.; Yang, H. "Lattice Strain Matching"-Enabled Nanocomposite Design to Harness the Exceptional Mechanical Properties of Nanomaterials in Bulk Forms. *Adv. Mater.* **2020**, *32*, 1904387. [CrossRef]
54. Romanova, V.; Balokhonov, R.; Emelianova, E.; Sinyakova, E.; Kazachenok, M. Early prediction of macroscale plastic strain localization in titanium from observation of mesoscale surface roughening. *Int. J. Mech. Sci.* **2019**, *161–162*, 105047. [CrossRef]

Article

Effect of Environmentally Friendly Oil on Ni-Ti Stent Wire Using Ultraprecision Magnetic Abrasive Finishing

Jeong Su Kim [1], Sung Sik Nam [2], Lida Heng [2], Byeong Sam Kim [3,*] and Sang Don Mun [1,2,*]

1 Department of Energy Storage/Conversion Engineering, Jeonbuk National University, 567, Baekje-daero, Deokjin-gu, Jeonju-si 54896, Korea; jeongsu1592@naver.com
2 Division of Mechanical Design Engineering, Jeonbuk National University, 567, Baekje-daero, Deokjin-gu, Jeonju-si 54896, Korea; nam6978@naver.com (S.S.N.); henglida1@gmail.com (L.H.)
3 Department of Smart Automotive Engineering, Wonkwang University, 460, Iksan-daero, Iksan-si 54538, Korea
* Correspondence: anvkbs@wku.ac.kr (B.S.K.); msd@jbnu.ac.kr (S.D.M.); Tel.: +82-63-850-6697 (B.S.K.); +82-63-270-4762 (S.D.M.)

Received: 27 August 2020; Accepted: 28 September 2020; Published: 30 September 2020

Abstract: Nickel-titanium (Ni-Ti) has been widely used to make shape-memory actuator wire for numerous medical industrial applications, with the result that it frequently comes into contact with the human body. High-quality and nontoxic surfaces of this material are therefore in high demand. We used a rotating magnetic field for an ultraprecision finishing of Ni-Ti stent wire biomaterials and evaluated the finishing technique's efficacy with different processing oils. To create nontoxic Ni-Ti stent wire, the industrial processing oils that are generally used in the surface improvement process were exchanged for oils with low environmental impacts, and processed under rotating magnetic fields at different speeds and processing times. The processing performance of the different oils was compared and verified. The results show that ultraprecision magnetic abrasive finishing that uses olive and castor oil improves surface roughness by 66.67%, and 45.83%, respectively. SEM and energy-dispersive X-ray spectroscopy (EDX) analyses of the finished components (before and after processing) showed that the material composition of the Ni-Ti stent wire was not changed. Additionally, the magnetic abrasive tool composition was not found on the surface of the finished Ni-Ti stent wire. In conclusion, the environmentally friendly oil effectively improved the diameter of the Ni-Ti stent wire, demonstrating the utility of olive and castor oil in ultraprecision finishing of Ni-Ti stent wire biomaterials.

Keywords: ultraprecision magnetic abrasive finishing (UPMAF); environmentally friendly oil; Ni-Ti stent wire; surface roughness (Ra); removed diameter (RD)

---

1. Introduction

Nickel-titanium (Ni-Ti) stent wire is a biomaterial, which have been widely used in various applications including medical devices [1–3]. For example, it is used for endovascular stents, which are useful in treating various heart diseases. Blood flow can be improved by inserting a collapsed nickel Ni-Ti stent into a vein and heating the wire, and it can serve as a substitute for sutures. As a result, high-quality surface finishes and mechanical functionality have become desirable characteristics of such biomaterials [4,5]. Conventional polishing or grinding can produce high-quality surfaces [6,7]. In previous works, many researchers have adopted some surface finishing methods for improving the surface accuracy of their products. Chang et al. [8] used the magnetic abrasive finishing process for improving a surface roughness (Ra) of cylindrical SKD11 materials using unbonded magnetic abrasive tools. According to his results, the Ra of cylindrical SKD11 materials was enhanced to 0.042 μm by the unbonded magnetic abrasive tools. Heng et al. [9] proposed a new manufacturing precision microdiameter ZrO_2 bar 800 μm in diameter using new magnetic pole designs via ultraprecision magnetic abrasive finishing. According to his results, a surface roughness Ra of ZrO_2 ceramic bar was enhanced to 0.02 μm within 40 s under the optimal conditions. Singh et al. [10] applied the magnetic abrasive finishing process for enhancing the accuracy of a plane workpiece with various important parameters (i.e., voltage (DC), finishing gap, rotating speed, and abrasive grain size). According to his study, the voltage and working gap are obtained to be the best parameters for a change in surface roughness (ΔRa). However, such techniques typically use industrial processing oils (e.g., light oil, oil mist, SEA oil), which contain toxic substances that are likely to exist on the finished surface [11], reducing the appeal of such products [12]. Park et al. [13] have explained that, in the machining process, when industrial oils such as petroleum-based oils are used, they possibly are more harmful to human health than ecological oils. Benedicto and Carou et al. [14] have reviewed the application of various machining fluids on the machining processes. The machining fluid have been widely used for reducing the machining temperature, and for removing microchips of metal workpieces. Despite these critical works, some disadvantages still remain, such as the environmental impact and health risks to workers. Li and Aghazadeh et al. [15] have studied the health effects associated with the toxicity from metalworking fluids (MWFs) in grinding processes. They have reported that a waste oil, coolant, or another lubricant can generate a toxicity characteristic leachate procedure, which could be effective in human health.

In recent years, a surface accuracy and dimensional accuracy of wire products have been improved by some advanced surface treatment methods (i.e., ion implantation and plasma coating) [16]. The ion implantation and plasma-coating technique have successfully improved the surface quality of some biomedical wire materials (i.e., Ni-Ti wire, TMA wire, beta-titanium wire, etc.) [17]. However, despite the potential advantages of these surface treatment methods, various limitations still exist. In the ion implantation process, some highly toxic gases (i.e., arsine (AsH_3), and phosphine (PH_3)) have been used [18]. Therefore, these toxic gases probably remain on the wire workpiece surface after processing by ion implantation. Rahman et al. [19] demonstrated that in the plasma coating process, the high voltage electrical shock has been supplied to the gap of electrodes for producing the plasma. Therefore, in order to overcome these problems, the environmentally friendly oils have been applied to the ultraprecision magnetic abrasive finishing (UPMAF) process for ultraprecision finishing of Ni-Ti stent wire. In this study, we evaluated the effectiveness of light oil, olive oil, and castor oil for the UPMAF process of Ni-Ti stent wire in terms of Ra, and removed diameter (RD). This research aims to elucidate the characteristics of Ni-Ti stent wire produced with an UPMAF process and a rotating magnetic field according to different processing oils. In addition, the effects of important input parameters (i.e., the rotating speed of the magnetic field, and different processing times) on Ra and removed diameter were studied in this research.

2. Experimental Method and Setup

A photograph and a schematic diagram of an UPMAF process using a rotating magnetic field for processing Ni-Ti stent wire are shown in Figures 1 and 2, respectively. The equipment comprised a magnetic abrasive finishing part, a spool-driving stepping motor, two driven spools, two fixing rollers, two proximity sensors, a sensor controller, an electric slider, a power supply, a programmable controller, an electrical slider controller, a stepping motor controller, and a stepping motor (speed range: 350–4000 rpm). To achieve a high efficiency of finishing accuracy of a Ni-Ti stent wire, two sets of Nd-Fe-B permanent magnets were used. The permanent magnets were composed of a south pole and north pole, which generated lines of magnetic force between the poles. The mixture for the magnetic abrasive particles consists of electrolytic iron particles, diamond abrasive particles, and processing oil. The abrasive particles were controlled by the magnetic force at room temperature (25 °C). To perform ultraprecision magnetic abrasive finishing, a Ni-Ti stent wire 120 mm in length was inserted inside the particulate brush of the magnetic abrasive tool and vibrated at 10 Hz. The finishing part rotated at up to 2000 rpm. The schematic view of magnetic force acting on magnetic abrasive particle during the UPMAF process is shown in Figure 3. As the Nd-Fe-B permanent magnets were used, 520-mT of magnetic flux density was obtained in the finishing zone. For the finishing mechanisms of an UPMAF process, "M" is a position, where a magnetic force, F_m, strongly pushes on the Fe particle. A magnetic force, F_m, is generated by the two forces, F_x on the x-component and F_y on the y-component. The force F_x acts on the magnetic abrasive particle along the direction of the line of the magnetic field. The force F_y is produced by the line of the magnetic field when the Ni-Ti wire material pushes out the bridges formed in the direction of magnetic equipotential lines. A magnetic force, F_m, acting on a magnetic abrasive particle can be expressed by Equations (1) and (2) [20].

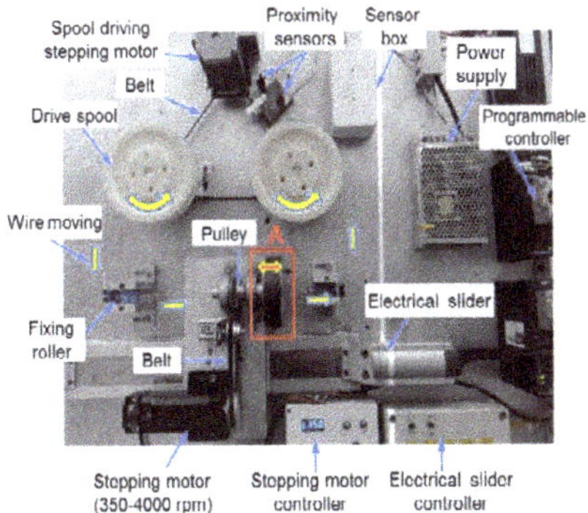

Figure 1. Photograph of ultraprecision magnetic abrasive finishing equipment for an ultraprecision Ni-Ti wire stent.

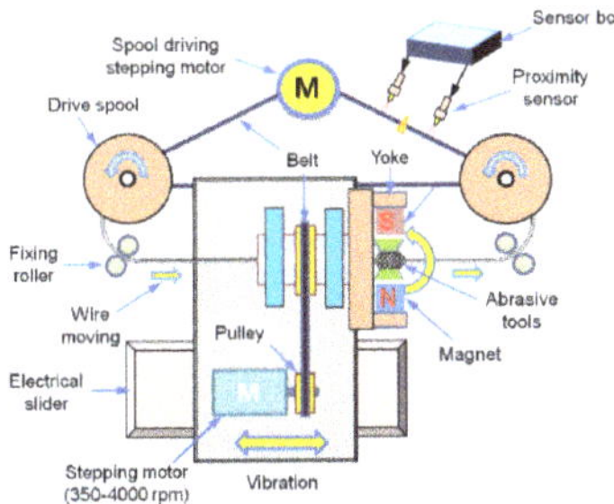

Figure 2. Diagram of ultraprecision magnetic abrasive finishing equipment for an ultraprecision of Ni-Ti wire.

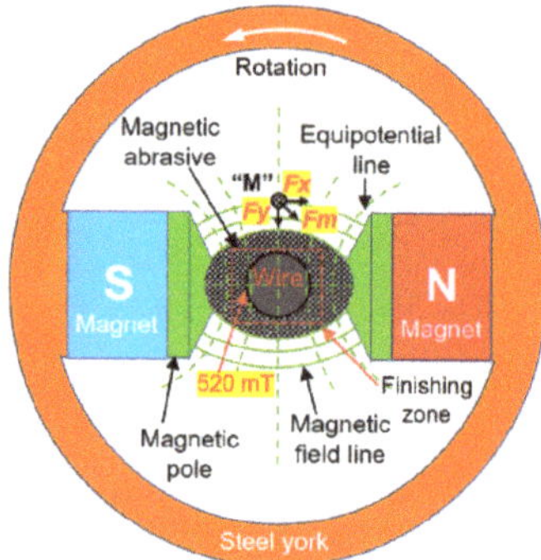

Figure 3. Schematic view of magnetic force acting on magnetic abrasive particle.

Where, x is direction of the line of magnetic field, y is direction of the magnetic equipotential line, χ_{FP} is the material magnetic susceptibility, μ is the permeability of free space, V is the volume of the magnetic abrasive particles, H is a magnetic field strength at point "M", $\frac{dH}{dX}$ and, $\frac{dH}{dY}$ are the variation rates of magnetic field strength in x and y components, respectively.

$$Fm = Fx + Fy \tag{1}$$

$$Fx = \chi_{\mathrm{FP}}\mu VH\left(\frac{dH}{dX}\right), \text{ and } Fy = \chi_{\mathrm{FP}}\mu VH\left(\frac{dH}{dY}\right) \tag{2}$$

Figure 4 is a photograph of the finishing part of an UPMAF process using a rotating magnetic field for the Ni-Ti stent wire. As shown in Figure 4, the Ni-Ti stent wire workpiece was put the gap between both magnetic poles, and surrounded by a mixture of unbonded magnetic abrasive particles. These magnetic abrasive particles were governed by magnetic forces. To perform an UPMAF process, a workpiece was moved inside the particulate brush of magnetic abrasive particles while the finishing part rotated. With this working procedure, the ultraprecision finishing of Ni-Ti stent wire was achieved.

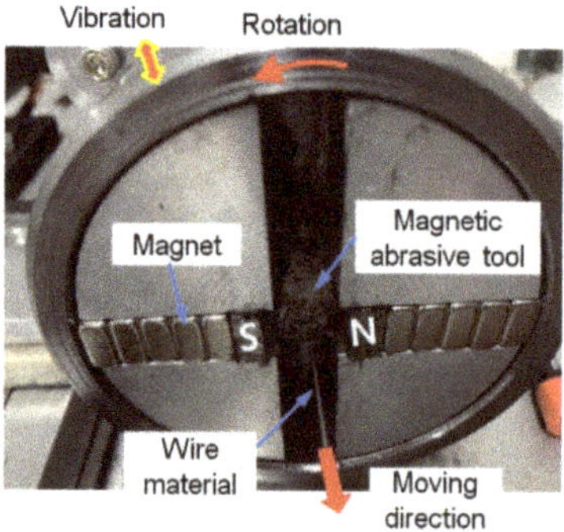

Figure 4. Photograph of the finishing part for ultraprecision finishing of a Ni-Ti wire stent.

2.1. Materials

In this study, the workpiece was Ni-Ti stent wire, a biomaterial commonly used in a variety of biomedical applications. The Ni-Ti stent wires were 120 mm in length and 0.5 mm in diameter, with an Ra of approximately 0.24 µm. A 2D dimension view of Ni-Ti stent wire workpiece is shown Figure 5. Figure 6 shows a photograph comparison of Ni-Ti stent wires used in this study. Tables 1 and 2 list the mechanical and chemical compositions of the Ni-Ti stent.

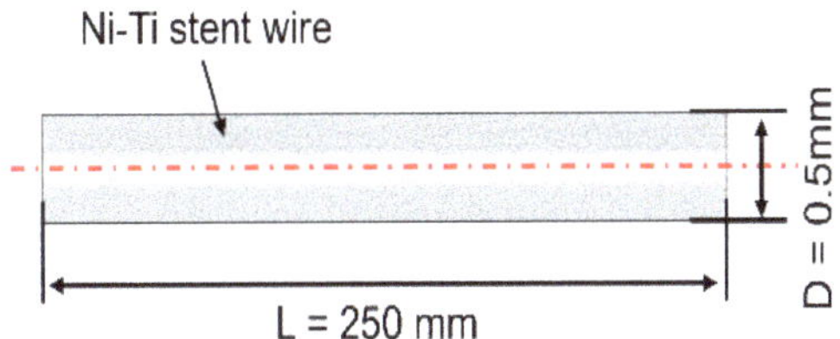

Figure 5. Two-dimensional view of Ni-Ti stent wire workpiece.

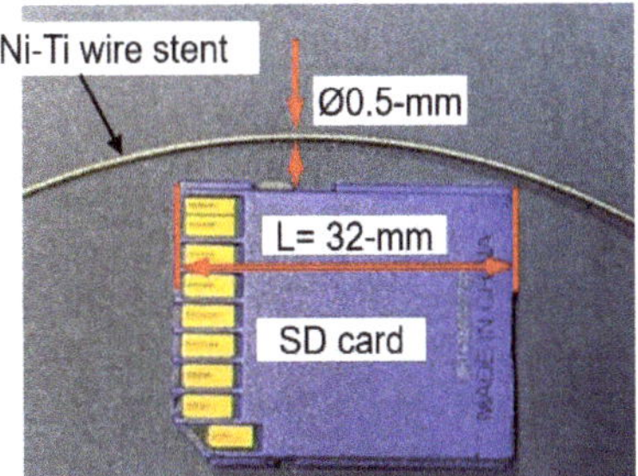

Figure 6. Photographic comparison of Ni-Ti stent wire material.

Table 1. Mechanical properties of Ni-Ti stent wire.

Properties	Value of Properties
Density	6.45 g/cm^3
Tensile strength	800–1500 MPa
Tensile yield strength	100–800 MPa
Poisson's ratio	0.33
Elastic modulus	70–110 GPa
Elongation at failure	1–20%

Table 2. Chemical composition of Ni-Ti wire.

Element	Spect.	Chemical Element (%)	Atomic
Ti K	ED	44.69	49.76
Ni K	ED	55.31	50.24
Total		100.00	100.00

2.2. Typical Processing Oils

Processing oils are essential elements of the finishing or grinding process, reducing the friction and high temperatures that occur between the surface finish of the sample and the abrasive particles [21]. During the UPMAF process, the mechanical friction between the relative motion of the sample and abrasive particles can cause microcracks on the finished surface. High temperatures generated during the finishing process can increase the wear of the abrasive tools, resulting in dimensional deviation and premature failure. To reduce both friction and temperatures in the finishing process, processing oils are applied to the mixtures of magnetic abrasive tools. In previous research, industrial processing oils have been used in the magnetic abrasive finishing process, but after the finishing process, undesirable toxic substances in the industrial oil are likely to exist on the finished surface, which can also suffer from unacceptable surface roughness. To overcome these problems, we replaced the industrial processing oils commonly used in finishing or machining processes with oil associated with low environmental impacts, including olive, castor, and light oil. A comparison of the properties of the processing oil used for UPMAF process is supplied in Table 3. Light oil has the lowest viscosity and density when compared with olive oil and castor oil, but the highest surface tension (31 dyne/cm), followed by castor oil and olive oil. However, castor oil has the highest viscosity among three oils.

Table 3. Characteristics of processing oil used in the finishing process.

Processing Oil	Temperature (°C)	Viscosity (Pa·s)	Density (g/cm^3)	Surface Tension (dyne/cm)
Castor oil	26	0.3115	0.956	14.89 ± 1.12
Olive oil	26	0.0341	0.857	10.00 ± 0.66
Light oil	26	0.005	0.8–0.82	31

2.3. Experimental Conditions

Detailed conditions for this experiment are supplied in Table 4. Ni-Ti stent wires with an Ra of 24 μm were chosen as the sample workpiece and finished at different rotating speeds of the magnetic field (i.e., 500, 1000, 1500, and 2000 rpm), for 150 s of total finishing time. Nd-Fe-B permanent magnets were utilized to generate the high magnetic force. The vibration frequency of the magnetic pole was 10 Hz with an amplitude of 5 mm, and the moving feed of the workpiece was 80 mm/min. Three different processing oils (light, olive, and castor oil) were applied during the process for comparison. Scanning electron microscope microimages were utilized to evaluate the changes in Ra of the Ni-Ti stent wire. For determination of surface roughness after processing, the average surface roughness (Ra) of Ni-Ti stent wire was measured at three different positions every 30 s of processing time by using a surface roughness tester (Mitutoyo SJ-400) (Mitutoyo, Sakado, Japan). Figure 7 shows a measuring procedure of surface roughness Ra value for wire material using a surface roughness tester (Mitutoyo SJ-400). As shown in the photo that during the measuring process, the tip was moved along with the length of wire material and the measuring length on wire sample was 5 mm with the measuring speed was 0.5 m/s. Also, the value of removed diameter of Ni-Ti stent wire was measured every 30 s of processing time by using a laser scan micrometer (Mitutoyo LSM-6200) (Mitutoyo, Sakado, Japan). The photos of the scanning electron microscope (SEM at 120×) were used to assess the effect of environmentally friendly oil on the improvement in Ra of the Ni-Ti stent wire in an UPMAF process using rotating magnetic field.

Table 4. Experimental conditions.

Workpiece material	Ni-Ti wire stent (L = 250 mm, D = 0.5 mm)
Electrolytic iron powder	0.8 g (Fe#200)
Diamond paste	0.5 μm (0.3 g)
Processing oil	0.2 mL (light oil, olive oil, castor oil)
Magnet type	Nd-Fe-B permanent magnet (size: 20 × 10 × 10 mm^3)
Magnetic pole shape	1 mm square edge
Amplitude	5 mm
Workpiece moving feed	80 mm/min
Rotational speed	500 rpm, 1000 rpm, 1500 rpm, 2000 rpm
Processing time	0 s, 30 s, 60 s, 90 s, 120 s, 150 s
Frequency	10 Hz
Magnetic flux density in finishing zone	520 mT
Working gap	2.25 mm

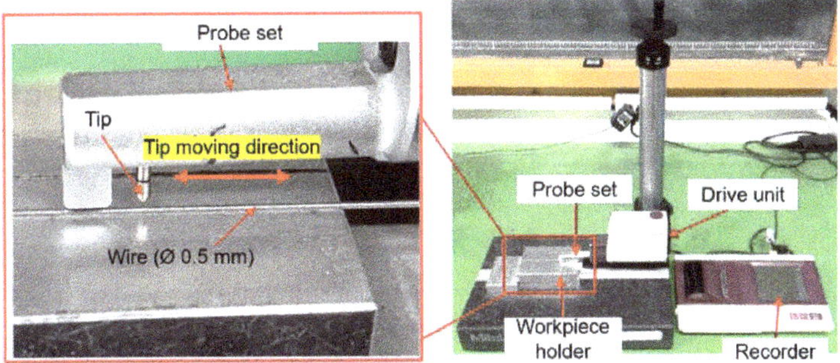

Figure 7. Measuring procedure of surface roughness value for wire material, using a surface roughness tester (Mitutoyo SJ-400).

3. Result and Discussion

To investigate the finishing characteristics of an ultraprecision magnetic abrasive finishing with different processing oils, the electrolytic iron powder (Fe#200) and diamond abrasive particles (0.5 μm) were mixed together with light oil, olive oil, and castor oil. The effect of different processing oils on finishing characteristics of Ni-Ti stent wires at different magnetic field rotating speeds was discussed.

3.1. Effect of Light Oil on Finishing Characteristics

To find the most optimal magnetic field rotating speed in terms of the surface roughness, the experiment was performed at a magnetic field rotating speed (500, 1000, 1500, and 2000 rpm), 10 Hz of vibration frequency, 5 mm of amplitude, and a feed rate of 80 mm/min. The effect of light oil on improvement in surface roughness of the Ni-Ti wire with various rotation speeds of the magnetic field at (500, 1000, 1500, 2000 rpm) is shown in Figure 8. As shown in Figure 8, Ra values of the Ni-Ti wire stent were significantly improved by light oil at all rotation speeds. The greatest improvement in Ra was obtained at 1500 rpm followed by 1000, 500, and 2000 rpm. This indicates that increasing the rotation speed of the magnetic field can improve Ra of an Ni-Ti wire stent. In the case of 1500 rpm, the Ra of the stent wire decreased from 0.24 μm to 0.07 μm over 150 s of processing time. However, at 2000 rpm of magnetic field rotating speed, the centrifugal force of the magnetic abrasive tools was increased, resulting in reduced the magnetic force, and therefore the magnetic abrasive tools flew in all directions.

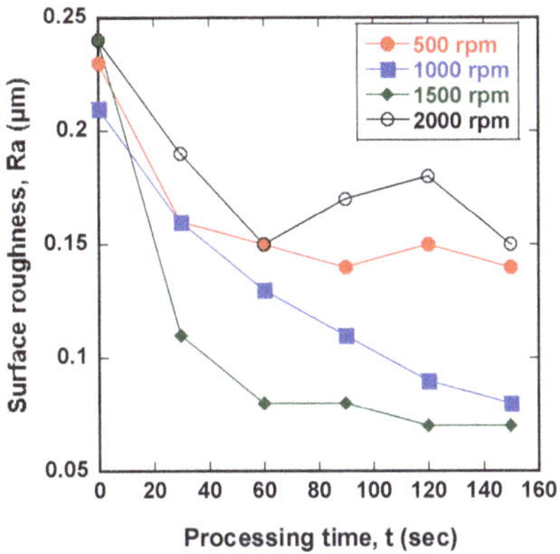

Figure 8. Correlation of surface roughness Ra vs. processing time, (light oil, diameter 0.5 μm, 80 mm/min).

3.2. Effect of Olive oil on Finishing Characteristics

Olive oil is an environmentally friendly substance that is generally utilized in the food and medical industries [22,23]. It has a viscosity of 0.0341 Pa·s, and a density of 0.857 kg/m^3. In this study, 0.2 mL of olive oil was combined with 0.8 g of electrolytic iron particles and 0.3 g of diamond paste at 25 °C using an unbonded magnetic abrasive method. The effect of olive oil on improvement in Ra of the stent wire at various rotating speeds of the magnetic field is shown in Figure 9. Olive oil significantly improved the surface roughness of Ni-Ti wire at all rotating speeds. However, 1500 rpm was found to be the optimal condition at 90 s of processing time. When 1500 rpm of rotating speed was used, the original Ra value of Ni-Ti improved from 0.24 μm to 0.07 μm for 90 s, after which the Ra did not improve further, because the unevenness of a surface of Ni-Ti stent wire was completely removed by that time. The slope of 2000 rpm shows worse improvement in Ra compared with the other conditions. The result can be attributed to the increase in the centrifugal force of magnetic abrasive tools.

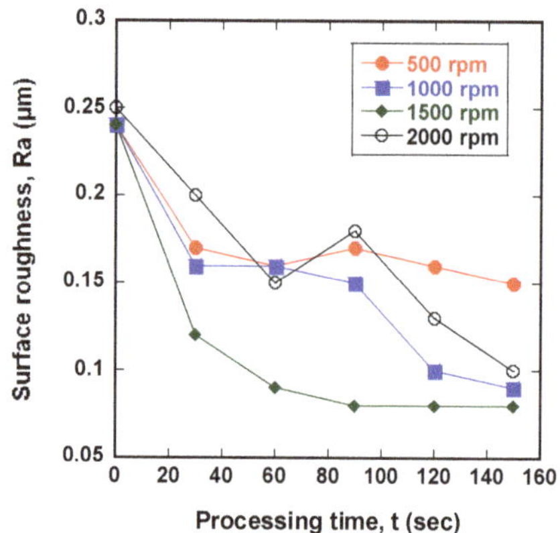

Figure 9. Correlation of surface roughness (Ra) vs. processing time, (olive oil, 0.5 μm, 80 mm/min).

3.3. Effect of Castor Oil on Finishing Characteristics

Castor oil is a colorless vegetable oil pressed from castor beans [24]. The boiling point of castor oil is 313 °C, its viscosity is 0.3115 Pa·s, and its density is 0.956 kg/m^3. This oil is commonly utilized in cosmetic products, including creams and moisturizers. In addition, it has been utilized to improve hair conditioning in other products due to supposed antidandruff properties. In this study, 0.2 mL of castor oil was combined with 0.8 g of electrolytic iron particles and 0.3g of diamond paste. The effect of castor oil on the improvement in Ra of the wire stent at various rotating speeds of the magnetic field (500, 1000, 1500, and 2000 rpm) is shown in Figure 10. As with the light oil and olive oil, the Ra of the Ni-Ti wire stent improved at all rotating speeds. The slope of 1500 rpm shows the greatest improvement in Ra, from 0.24 μm to 0.16 μm for 90 s of processing time.

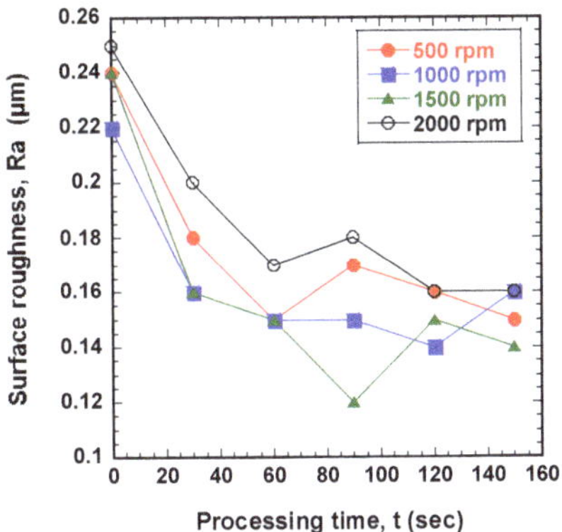

Figure 10. Correlation of surface roughness (Ra) vs. processing time, (castor oil, 0.5 µm, 80 mm/min).

3.4. *Percentage Improvement in Surface Roughness (PIISR)*

The effect of different processing oils on the improvement in surface roughness of Ni-Ti stent wire is shown in Figure 11. The processing time of each workpiece was 150 s. In the bar graph, the surface roughness of the Ni-Ti wire workpiece rapidly improves from 0 s to 30 s of processing time, and then improves at a diminished rate until 150 s. The greatest improvement in Ra was obtained with light oil, followed by olive oil, and castor oil. In the case of industrial oil, the Ra value decreased from 0.24 µm to 0.07 µm. With olive and castor oils, Ra values improved from 0.24 µm to 0.08 µm and from 0.24 µm to 0.12 µm, respectively. The percentage of improvement formula for Ni-Ti wire stent Ra as a function of different processing oils can be expressed by a formula. ***BUPMAF*** (before ultraprecision magnetic abrasive finishing) is the Ra value before processing and ***AUPMAF*** (after ultraprecision magnetic abrasive finishing) is the value after 150 s of processing. *PIISR* is the percentage improvement in surface roughness, expressed as the rate of change of Ra for each set of processing conditions as a percentage of the improvement in Ra.

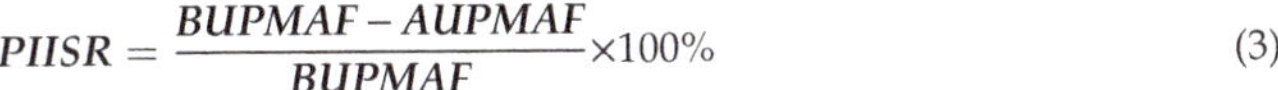

$$PIISR = \frac{BUPMAF - AUPMAF}{BUPMAF} \times 100\% \quad (3)$$

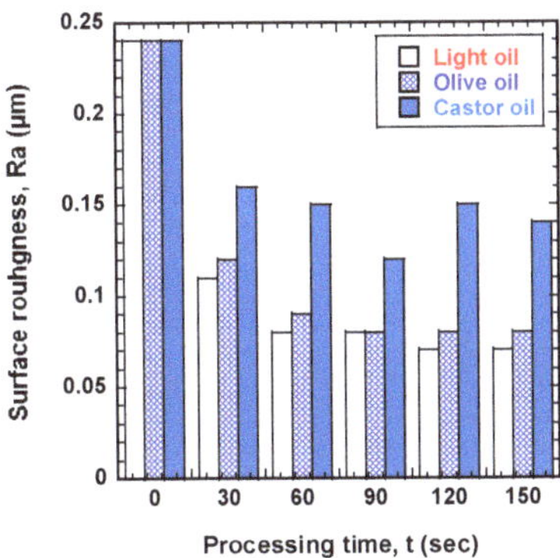

Figure 11. Correlation of surface roughness Ra vs. processing time according to processing oils, (1500 rpm, 0.5 µm, 80 mm/min).

A *PIISR* graph of the improvement in surface roughness Ra before and after processing according to processing oil within 150 s of processing time is shown in Figure 12. The results show that improvement in Ra for each processing oil was greater than 45%. Light oil was associated with an improvement of 70.83%, while olive and castor oil resulted in improvements of 66.67% and 45.833%, respectively. Light oil is therefore the preferred oil, followed by olive oil and castor oil. This result can be attributed to the viscosity of the processing oils. As shown in Table 3, the viscosity values of light oil, olive oil, and castor oil 0.3115, 0.0341, and 0.005, respectively. From this it can be concluded that low-viscosity oil produced greater improvements in Ra compared with oils of higher viscosity. Light oil, which has the lowest viscosity and density, reduced the temperature and friction generated during finishing and resulted in a high-quality Ra value. According to the lowest viscosity of light oil (0.005 Pa·s), the Ra value obtained by this condition should be much better than olive oil. However, the Ra value is difficult to improve to 0.07 μm. This is probably due to the effect of another parameter on Ra improvement, such as 0.5 μm of abrasive, which cannot enhance the Ra value less than 0.07 μm.

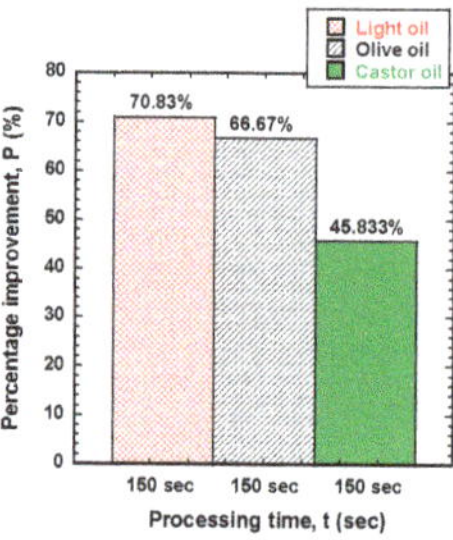

Figure 12. Correlation of percentage improvement in surface roughness (*PIISR*) vs. processing time under optimal conditons, (1500 rpm, 150 s, 0.5 μm 10 Hz).

We can conclude that environmentally friendly processing oils can be used for precision finishing of Ni-Ti wire stents via an UPMAF process. Olive oil was chosen as the processing oil for precision finishing of Ni-Ti wire material at 500, 1000, 1500, and 2000 rpm of a rotating magnetic field for 150 s. The effect of olive oil on the removed diameter of Ni-Ti against processing time is shown in Figure 13. The results show that the removed diameter (RD) of Ni-Ti wires can be significantly increased in all conditions of a rotating magnetic field. In terms of the RD, the diameters of the wire materials removed at 500, 1000, 1500, and 2000 rpm were 0.00130, 0.00212, 0.00281, and 0.00192 mm, respectively. As with Ra improvement, 1500 rpm was associated with the largest reduction in diameter. This can be explained by the fact that, at 2000 rpm of rotational speed, the unbonded magnetic abrasive particles have enough time for removing the irregular scratches from the Ni-Ti stent wire's surface. Therefore, it can be confirmed that, when increasing the rotating speed to a certain level, the best result can be received.

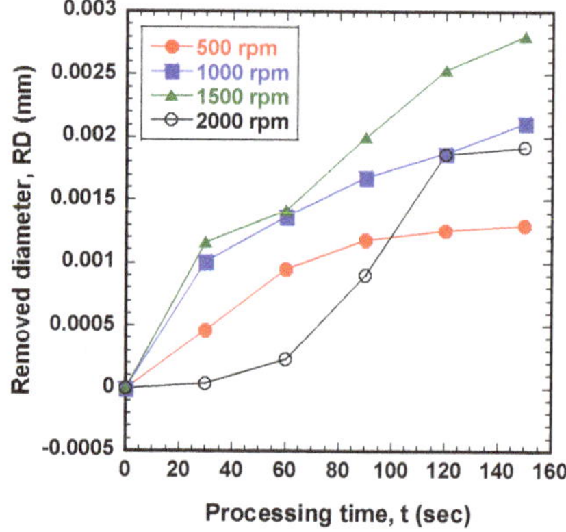

Figure 13. Correlation of removed diameter vs. processing time (olive oil, 0.5 μm, 80 mm/min).

SEM microimages of Ni-Ti stent wires before and after processing by an UPMAF process (magnified 120 times) are shown in Figure 14. The surface conditions of Ni-Ti stent wires before finishing are shown in Figure 14a. Initial scratches and unevenness were found throughout the surface of the stent wire, and the initial Ra was 0.24 µm. The surface condition of the wire after processing by light oil, olive oil, and castor oil can be found in the Figure 14b–d), respectively. It was confirmed that the finished surfaces of the Ni-Ti wire were smoother than the surface before processing. The surfaces finished with light oil and olive oil were significantly smoother than before processing, with Ra values of 0.07 µm and 0.08 µm, respectively. However, the surfaces finished with castor oil were not as smooth as those finished with light oil and olive oil. As can be seen in Figure 14d, the original scratches and irregular asperities remain on the surface. Energy-dispersive X-ray spectroscopy (EDX) analysis of a wire stent produced by ultraprecision magnetic abrasive finishing with olive oil is shown in Figure 15. The chemical composition of the stent wire before processing is shown in Figure 15a. The analysis result of EDX test shows that 44.60% Ti and 55.40% Ni were detected at the Ni-Ti stent wire surface. After processing with olive oil, 44.69% Ti, and 55.31% Ni were detected at the surface, as seen in Figure 15b. In addition, EDX analysis revealed no toxic substances on the finished surfaces of the Ni-Ti wire.

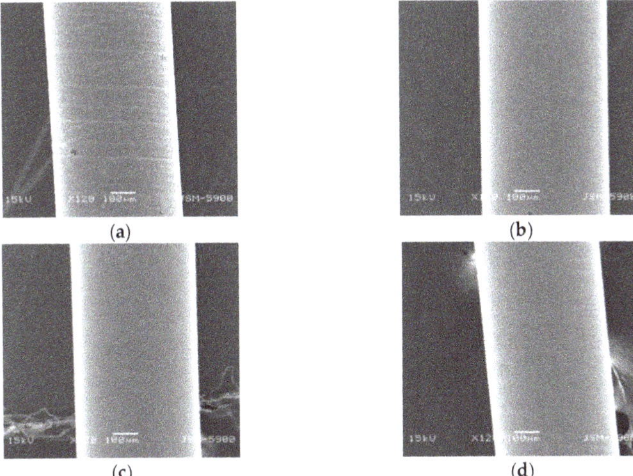

Figure 14. Surface of the workpiece before and after finishing. (**a**) Before finishing, Ra = 0.24 µm; (**b**) Finished with light oil, Ra = 0.07 µm; (**c**) Finished with olive oil Ra = 0.08 µm; (**d**) Finished with castor oil Ra = 0.12 µm.

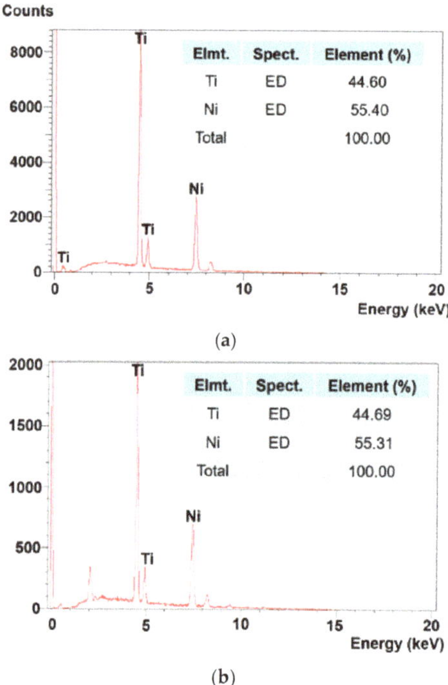

Figure 15. Energy-dispersive X-ray spectroscopy (EDX) test results of Ni-Ti wire before and after processing by ultraprecision magnetic abrasive. (**a**) Components of the workpiece before processing (Ra: 0.24 μm, processing time: 0 s); (**b**) Components of the workpiece after processing (light oil, Ra: 0.08 μm, processing time: 120 s).

4. Conclusions

In this research, an UPMAF process using rotating magnetic field was utilized to the high-precision finishing of Ni-Ti stent wire material. Vegetable oil was used as the finishing oil and was compared with the industrial finishing oil that is commonly utilized in the conventional magnetic abrasive finishing process. The characteristics and finishing abilities of these techniques were compared, and the Ni-Ti stent wire workpiece was finished by the different rotational speeds of the magnetic field with vegetable oil as finishing oil, and the following results were shown:

1. To study the characteristics of different processing oils, two types of vegetable oil (olive oil and castor oil) were used and compared with industrial light oil. The improvements in Ra with light oil, olive oil, and castor oil were equivalent to treatment at 0.07 μm, 0.08 μm, and 0.12 μm, respectively. In the cases of light oil and olive oil, the deviation in improvement was not significantly different. Therefore, the industrial oil that has been widely used in the finishing process for Ni-Ti wire can be replaced by olive oil.
2. Light oil, olive oil, and castor oil improved Ra values by 70.83%, 66.67%, and 45.83%, respectively. In each case the improvement was greater than 45.83%. The different results of improvement in surface roughness Ra can be explained based on the different value of the finishing oil's viscosity. As the lowest viscosity is that of light oil (0.005 Pa·s), the Ra value obtained by this condition should be much better than olive oil. However, the Ra value is difficult to improve to 0.07 μm. This is probably due to the effect of another parameter such as 0.5 μm of abrasive on Ra improvement, which cannot enhance the Ra value less than 0.07 μm.

3. In terms of the removed diameter, environmentally friendly oil can reduce the diameter of Ni-Ti stent wires by 0.00281 mm after 150 s while improving the surface roughness Ra from 0.24 μm to 0.08 μm. This can be confirmed that an ultraprecision magnetic abrasive finishing process with environmentally friendly oil can reduce the surface roughness value and diameter of Ni-Ti stent wire simultaneously.
4. Olive oil exhibited excellent performance in terms of Ra and removed diameter. An EDX analysis found no components of the processing oil and or toxic substances on the surface finish of the Ni-Ti stent wire, indicating that olive oil can be applied to ultraprecision surface finishing.

Author Contributions: Design experiment, methodology and conceptualization J.S.K. and L.H.; investigation and editing, S.S.N., B.S.K. and S.D.M.; writing of paper, J.S.K. All authors have read and agreed to the published version of the manuscript.

Funding: This research was funded by NATIONAL RESEARCH FOUNDATION (NRF) of Korea in 2019, (Research Project No.2016R1D1A1B03932103, 2019R1F1A1061819).

References

1. Guo, Y.; Klink, A.; Fu, C.; Snyder, J. Machinability and surface integrity of Nitinol shape memory alloy. *Cirp Ann.* **2013**, *62*, 83–86. [CrossRef]
2. Duerig, T.; Pelton, A.; Stöckel, D. An overview of nitinol medical applications. *Mater. Sci. Eng. A* **1999**, *273*, 149–160. [CrossRef]
3. Shabalovskaya, S.; Anderegg, J.; Van Humbeeck, J. Critical overview of Nitinol surfaces and their modifications for medical applications. *Acta Biomater.* **2008**, *4*, 447–467. [CrossRef] [PubMed]
4. Shabalovskaya, S.; Rondelli, G.; Rettenmayr, M. Nitinol surfaces for implantation. *J. Mater. Eng. Perform.* **2009**, *18*, 470–474. [CrossRef]
5. Hassel, A.W. Surface treatment of NiTi for medical applications. *Minim. Invasive Ther. Allied Technol.* **2004**, *13*, 240–247. [CrossRef]
6. Heng, L.; Kim, Y.J.; Mun, S.D. Review of Superfinishing by the Magnetic Abrasive Finishing Process. *High Speed Mach.* **2017**, *3*, 42–55. [CrossRef]
7. Heng, L.; Yin, C.; Han, S.H.; Song, J.H.; Mun, S.D. Development of a New Ultra-High-Precision Magnetic Abrasive Finishing for Wire Material Using a Rotating Magnetic Field. *Materials* **2019**, *12*, 312. [CrossRef]
8. Chang, G.W.; Yan, B.H.; Hsu, R.T. Study on cylindrical magnetic abrasive finishing using unbonded magnetic abrasives. *Int. J. Mach. Tools Manuf.* **2002**, *42*, 575–583. [CrossRef]
9. Heng, L.; Kim, J.S.; Tu, J.F.; Mun, S.D. Fabrication of precision meso-scale diameter ZrO2 ceramic bars using new magnetic pole designs in ultra-precision magnetic abrasive finishing. *Ceram. Int.* 2020. [CrossRef]
10. Singh, D.K.; Jain, V.K.; Raghuram, V. Parametric study of magnetic abrasive finishing process. *J. Mater. Process. Technol.* **2004**, *149*, 22–29. [CrossRef]
11. Yin, C.; Heng, L.; Kim, J.; Kim, M.; Mun, S. Development of a New Ecological Magnetic Abrasive Tool for Finishing Bio-Wire Material. *Materials* **2019**, *12*, 714. [CrossRef] [PubMed]
12. Amini, S.; Baraheni, M.; Esmaeili, S.J. Experimental comparison of MO40 steel surface grinding performance under different cooling techniques. *Int. J. Lightweight Mater. Manuf.* **2019**, *2*, 330–337. [CrossRef]
13. Park, K.H.; Olortegui-Yume, J.; Yoon, M.C.; Kwon, P. A study on droplets and their distribution for minimum quantity lubrication (MQL). *Int. J. Mach. Tools Manuf.* **2010**, *50*, 824–833. [CrossRef]
14. Benedicto, E.; Carou, D.; Rubio, E.M. Technical, economic and environmental review of the lubrication/cooling systems used in machining processes. *Procedia Eng.* **2017**, *184*, 99–116. [CrossRef]
15. Li, K.; Aghazadeh, F.; Hatipkarasulu, S.; Ray, T.G. Health risks from exposure to metal-working fluids in machining and grinding operations. *Int. J. Occup. Saf. Ergon.* **2003**, *9*, 75–95. [CrossRef] [PubMed]
16. Jabbari, Y.S.A.; Fehrman, J.; Barnes, A.C.; Zapf, A.M.; Zinelis, S.; Berzins, D.W. Titanium nitride and nitrogen ion implanted coated dental materials. *Coatings* **2012**, *2*, 160–178. [CrossRef]
17. Krishnan, M.; Saraswathy, S.; Sukumaran, K.; Abraham, K.M. Effect of ion-implantation on surface characteristics of nickel titanium and titanium molybdenum alloy arch wires. *Indian J. Dent. Res.* **2013**, *24*, 411–417. [CrossRef]

18. Vavilov, V.S. Possibilities and limitations of ion implantation in diamond, and comparison with other doping methods. *Physics-Uspekhi* **1994**, *37*, 407–411. [CrossRef]
19. Rahman, M.; Haider, J.; Hashmi, M.J. Health and Safety Issues in Emerging Surface Engineering Techniques. *Compr. Mater. Process.* **2014**, *8*, 35–47.
20. Djavanroodi, F. Artificial neural network modeling of surface roughness in magnetic abrasive finishing process. *Res. J. Appl. Sci. Eng. Technol.* **2013**, *6*, 1976–1983. [CrossRef]
21. Yara-Varón, E.; Li, Y.; Balcells, M.; Canela-Garayoa, R.; Fabiano-Tixier, A.S.; Chemat, F. Vegetable oils as alternative solvents for green oleo-extraction, purification and formulation of food and natural products. *Molecules* **2017**, *22*, 1474. [CrossRef] [PubMed]
22. Kala, P.; Pandey, P.M. Comparison of finishing characteristics of two paramagnetic materials using double disc magnetic abrasive finishing. *J. Manuf. Process.* **2015**, *17*, 63–77. [CrossRef]
23. Jain, V.K. Magnetic field assisted abrasive based micro-/nano-finishing. *J. Mater. Process. Technol.* **2009**, *209*, 6022–6038. [CrossRef]
24. Alfred, T.; Matthäus, B.; Fiebig, H.J. Fats and fatty oils. In *Ullmann's Encyclopedia of Industrial Chemistry*; Wiley-VCH: Weinheim, Germany, 2000; pp. 1–84.

Article

Fabrication of a Novel Ta(Zn)O Thin Film on Titanium by Magnetron Sputtering and Plasma Electrolytic Oxidation for Cell Biocompatibilities and Antibacterial Applications

Heng-Li Huang [1,2,†], Ming-Tzu Tsai [3,†], Yin-Yu Chang [4,*], Yi-Jyun Lin [4] and Jui-Ting Hsu [1,2]

1 School of Dentistry, China Medical University, Taichung 404, Taiwan; henleyh@gmail.com (H.-L.H.); jthsu@mail.cmu.edu.tw (J.-T.H.)
2 Department of Bioinformatics and Medical Engineering, Asia University, Taichung 413, Taiwan
3 Department of Biomedical Engineering, Hungkuang University, Taichung 433, Taiwan; anniemtt@sunrise.hk.edu.tw
4 Department of Mechanical and Computer-Aided Engineering, National Formosa University, Yunlin 632, Taiwan; xyz10920@yahoo.com.tw
* Correspondence: yinyu@gs.nfu.edu.tw; Tel.: +1-886-5-631-355
† Equal contributors.

Received: 29 April 2020; Accepted: 16 May 2020; Published: 18 May 2020

Abstract: Pure titanium (Ti) and titanium alloys are widely used as artificial implant materials for biomedical applications. The excellent biocompatibility of Ti has been attributed to the presence of a natural or artificial surface layer of titanium dioxide. Zinc oxide and tantalum oxide thin films are recognized due to their outstanding antibacterial properties. In this study, high power impulse magnetron sputtering (HiPIMS) was used for the deposition of tantalum oxide and zinc-doped Ta(Zn)O thin films on Ti with rough and porous surface, which was pretreated by plasma electrolytic oxidation (PEO). Surface morphology, antibacterial property as well as cell biocompatibility were analyzed. The antibacterial effect was studied individually for the Gram-positive and Gram-negative bacteria *Staphylococcus aureus* (*S. aureus*) and *Actinobacillus actinomycetemcomitans* (*A. actinomycetemcomitans*). The deposited Ta (Zn)O coating was composed of amorphous tantalum oxide and crystalline ZnO. The antibacterial results on the tantalum oxide and Ta(Zn)O coated Ti indicated a significant inhibition of both *S. aureus* and *A. actinomycetemcomitans* bacteria when compared with the uncoated Ti samples. The deposited Ta(Zn)O showed the best antibacterial performance. The Ta(Zn)O coated Ti showed lower level of the cell viability in MG-63 cells compared to other groups, indicating that Zn-doped Ta(Zn)O coatings may restrict the cell viability of hard tissue-derived MG-63 cells. However, the biocompatibility tests demonstrated that the tantalum oxide and Ta(Zn)O coatings improved cell attachment and cell growth in human skin fibroblasts. The cytotoxicity was found similar between the Ta_2O_5 and Ta(Zn)O coated Ti. By adopting a first PEO surface modification and a subsequent HiPIMS coating deposition, we synthetized amorphous tantalum oxide and Ta(Zn)O coatings that improved titanium surface properties and morphologies, making them a good surface treatment for titanium-based implants.

Keywords: high power impulse magnetron sputtering; zinc oxide; tantalum oxide; thin film; plasma electrolytic oxidation; antibacterial; biocompatibility

1. Introduction

With the growing demand for metal implants in the medical industry, surface treatment, which includes various types of bio-coating and different surface modification techniques used for fabrication of biomaterials, is widely applied in implants for enhancing the biocompatibility and antibacterial properties between metals and tissues. This method can improve the biocompatibility of the metal implant surface and preserve the mechanical properties of the implant body [1–3]. Although dental and orthopedic implants have improved their biomedical properties and enhanced the life of many patients, they are not free from the influence of bacteria. Pure titanium and its alloys are the most commonly used materials for permanent implants in contact with soft and hard tissues. Microbial infection is one of the main causes of implant failure. Most of these infections are caused by common bacteria, such as *Staphylococcus aureus* (*S. aureus*) and *Actinobacillus actinomycetemocomitans* (*A. actinomycetemocomitans*) which show enhanced activity on metallic and biocompatible surfaces used for implants [4,5]. Therefore, it is worth using surface treatments with two main categories: surface modification (physical, chemical or combined) and coatings (physical, chemical or combined).

Different surface modification methods are used to improve mechanical, tribological and biomedical properties of titanium include plasma nitriding, electrodeposition and anodizing, etc. Even though each treatment has its own unique advantage, plasma electrolytic oxidation (PEO) treatment, also called anodic spark oxidation or micro-arc oxidation, has become increasingly popular for titanium alloys owing to the excellent adhesion strength of the oxides on the surface and environmental friendliness [6]. This method produces a bioactive surface with a rough and porous structure at the surface of a work piece electrode immersed in an appropriate electrolyte by applying a high pulsed voltage [7]. For example, with respect to the PEO process, a bioactive TiO_2 layer can be formed on the surface of Ti. In addition, many studies have been directed towards making the surfaces of implants as biocompatible and resistant against bacterial attachment as possible, particularly for titanium-based alloys which are often the materials of choice for dental and orthopedic implants. For example, titanium surfaces have been modified by photocatalytic titanium dioxide (TiO_2) coatings and alloying titanium with silver and copper or producing the TiO_2 surface layer with silver or zinc [8,9]. Tantalum (Ta) is a transition element with substantial chemical inertness under room temperature [10,11]. Ta is also a biocompatible metal and can induce osseointegration between bone tissue and it surface [12]. In clinics, porous Ta implants have been used widely in orthopedics and dentistry owing to their excellent biocompatibilities [13–15]. For tantalum oxide materials, our pervious study [16] also revealed that Ta_2O_5 films are antibacterial and possess considerable biocompatibility with fibroblast cells. Furthermore, Ta_2O_5 films are more resistant to corrosion than TiO_2 films [17]. Besides, zinc (Zn) is an essential microelement involved with the different types of metabolism in the human body [18]. An appropriate amount of Zn can improve the integration of bone protein, stimulate osteoblast differentiation, and further increase the proliferation of bone cells [19]. Concerning the positive correlation between alkaline phosphatase (ALP) and osteoblast differentiation, a previous research by Yamaguchi et al. [20] indicated that a boost in ALP signal takes place when Zn is added to a human bone. The research conducted by Jin et al. [21] demonstrated that after implanting Zn into the surface of Ti, the antibacterial property was increased. ZnO was revealed to have antibacterial property against *Staphylococcus aureus* and *Escherichia coli* [22–24]. A duplex treatment combining PEO and physical vapor deposition (PVD) has been reported as enhancing mechanical properties and antibacterial performance [25]. The study proved that the Ag-contained TiO_2 coatings synthesized by combining PEO and magnetron sputtering possessed improved antibacterial activities without cytotoxic effect. Huang et al. also showed that the duplex treatment consisting of PEO and PVD could synthesize porous Ta_2O_5 films to have good biocompatibility [26].

Taking into account all of the above, the combination of modified Ta_2O_5 films possessing bactericidal effects with rough and porous Ti surfaces that inhibit bacterial adhesion and keep cell viability can be a promising strategy in order to create a synergic interaction that enhances the antibacterial properties and cell biocompatibilities of the final coating. The purpose of this study is to synthesize zinc-doped Ta(Zn)O films to form a bioactive and antibacterial surface which maybe more helpful for the possibility of clinical applications. Among the PVD technologies, HiPIMS is a comparatively new magnetron sputtering technique that provides a higher plasma ionization ability for the deposition of higher quality coatings with dense microstructure. Reactive-HiPIMS technique utilizes the magnetron sputtering in a reactive gas environment. It brings possibilities to enhance and tailor coating properties and often possesses better adhesion to the substrate, especially for those of oxide coatings, such as photoactive and antibacterial copper oxides and TiO_2.

This study applied plasma electrolytic oxidation (PEO) to pure Ti specimens to form rough and porous TiO_2 surfaces for biocompatible purpose and further used HiPIMS in a reactive environment with oxygen to deposit zinc-doped Ta(Zn)O films onto the surface of the PEO-treated Ti. This study investigated the antibacterial property of the coating against Gram-positive and Gram-negative bacteria. ISO 10993-5 cytotoxicity analyses were performed in soft and hard tissues, and MTT assays were applied to assess the cell survival rate.

2. Materials and Methods

2.1. Sample Preparation

Pure Ti samples (ASTM B265 Grade 2) were used as the experimental starting material. In order to remove the surface containment, the Ti samples were ground using abrasive papers and polished with colloidal alumina suspension to obtain the surface roughness Ra~0.5 μm, and then the samples were rinsed ultrasonically in alcohol solution for 15 min and dried in hot air. Sand blasting of the pure Ti was then conducted to obtain high adhesion strength for the following PEO treatment and deposition of tantalum oxide and Ta(Zn)O films. Figure 1 shows the illustration of the technical processing route of the samples. After sand blasting, all the Ti samples were ultrasonically cleaned with a successive ethyl alcohol, and distilled water; and then air-dried. PEO was used to produce rough and porous oxide films on the Ti surfaces. An alternating bipolar pulse power source was used to fabricate the oxide films. The pulse parameters including voltage, current, frequency, and duty cycle can be adjusted independently. In this study, the PEO was operated using a pulsed electrical signal (500 Hz, +280 V/−15 V and 50% duty cycle) and a current density limit of 120 mA/cm^2. The Ti sample was set as an anode, and a flat stainless steel was used as the cathode. A mixed aqueous solution containing 0.15 M $Ca(CH_3COO)_2{\cdot}H_2O$ and 0.0 2M $C_3H_7CaO_6P{\cdot}H_2O$ was used as the electrolyte. The PEO procedure was conducted for 90 s. The electrolyte was cooled by a cooling system to keep its temperature below 50 °C. After PEO treatment, the samples were taken out of electrolyte, ultrasonically cleaned in ethyl alcohol and air-dried before performing the following deposition of tantalum oxide and Ta(Zn)O films.

Tantalum oxide and Ta(Zn)O films were deposited on the PEO pretreated Ti by using HiPIMS with high-purity Ta and Zn targets (99.99 at.%, dia. = 7.5 mm). A pulsed power generator (TruPlasma Highpulse Series 4000, TRUMPF Hüttinger GmbH Co., Ditzingen, Germany) was connected to the targets. The Ti samples were placed on a substrate holder for the deposition, and the distance between the target and Ti substrates was 60 mm. A base pressure prior to deposition was less than 1.3×10^{-3} Pa. Before deposition, the substrates were ion etched for 20 min at a substrate bias potential of −800 V in argon (Ar) plasma. To enhance film adhesion, Ta ion bombardment using bias voltage of −800 V was applied before the deposition. For the deposition of tantalum oxide and Ta(Zn)O films, the flow rate of Ar was set at 50 sccm and oxygen (O_2) was introduced through a separate mass flow controller to maintain a deposition pressure of 0.4 Pa for the reactive sputtering deposition. In this study, a unipolar mode was used with a fixed pulse on-time of 20 μs, frequency of 500 Hz and duty cycle of 5%. A bias voltage of −60 V in the negative periods was connected to the Ti substrates during deposition. The Ta(Zn)O films were deposited with the cathode powers of the Ta target and Zn target set to 500 and 200 W, respectively. All the deposition experiments were performed without additional heating, and the substrate temperature was all below 80 °C. The total thickness of the coatings was controlled to 0.2~0.3 μm by using a deposition time of 15 min.

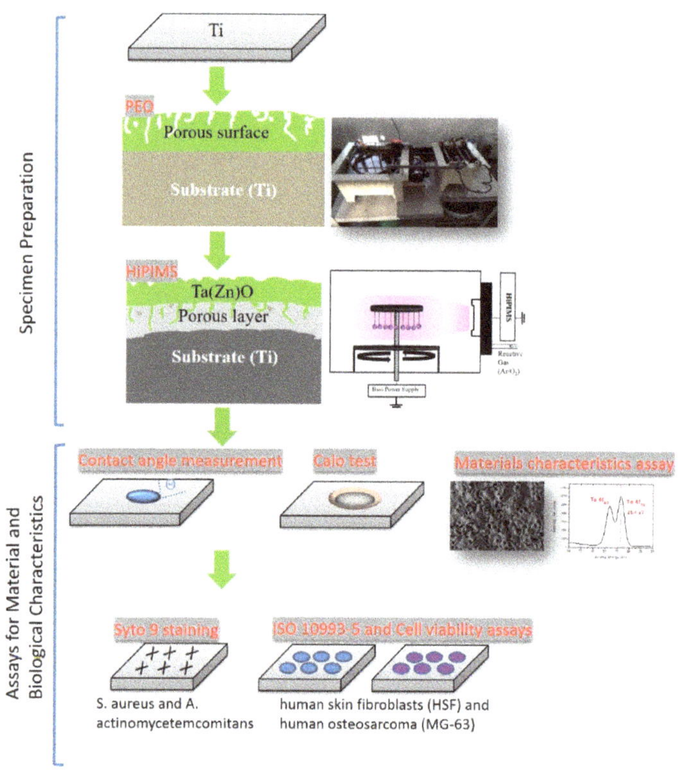

Figure 1. Schematic illustration of the technical processing route.

2.2. Materials Characteristics Measurement

The surface morphology and chemical composition of the coated samples were investigated by using a JSM-6700F field emission scanning electron microscope (FESEM, JEOL Inc., Tokyo, Japan), which equipped with an energy-dispersive x-ray spectroscopy (EDS) system. Optical microscope instruments are preferred for the evaluation of surface morphology of soft Ti implant materials. An accurate measure of both the magnitude and complexity of the surface extensions beyond the mean plane is necessary. In this study, the surface morphology and surface roughness were examined with a 3D laser scanning microscope (VK-X100, Keyence Inc., Osaka, Japan). A typical surface roughness Ra value, which showed the arithmetic mean value of the surface differences from the surface, was measured. In addition, the point height of irregularities Rz, which showed the average value of the absolute heights of five highest profile peaks and the depths of five deepest valleys within the evaluated length, was measured. An X-ray photoelectron spectroscope (PHI1600 XPS, Physical Electronics Inc., Chanhassen, MN, USA) with non-monochromatized Mg Kα radiation was used to identify the chemical binding of the tantalum oxide and zinc-doped Ta(Zn)O coatings. The sample surface was cleaned for 60 s using an Ar-ion gun to eliminate the influence of the surface contamination as much as possible before the spectrum was collected. Energy calibration was conducted by reference to the Au $4f_{7/2}$ peak at 83.8 eV from a clean gold surface. The spectral ranges at 26 ± 12 eV, 1035 ± 20 eV and 531 ± 10 eV, corresponded to the binding energies of Ta4f, Zn2p and O1s, respectively. X-ray diffraction (XRD) patterns were obtained with a D8 series X-ray diffractometer (Bruker Taiwan Co., Ltd., New Taipei City, Taiwan) using the CuKα radiations (λ = 1.54059 Å) in the 2θ range between 20° and 80° with a scanning step size of 0.02°. It was performed at a low glancing incidence angle of 2°

allowing identifying the coating structure. A Calo test, also known as a ball cratering test, was used to measure the thickness of the deposited coatings. The Calo tester consisted of a holder for the coated surface to be tested and a steel ball that was rotated against the measured surface by a rotating shaft. An optical microscope was used to take measurements of the crater diameter, and the coating thickness was calculated and obtained.

2.3. Contact Angle Measurement

After each coating sample was washed for 30 min alternately into containers with ethanol and deionized water in an ultrasonic cleaner, samples were dried in a clean and dry oven at 55 °C for 6 h. Then deionized water dropped from a micrometric syringe and touched onto the surface of untreated Ti (control), Ta_2O_5 and Ta(Zn)O coated samples. The status of the water on the samples was photographed by an optical instrument to measure the static contact angles of water on all samples by the measurement device (FTA-125, First Ten Angstroms, Portsmouth, VA, USA) at room temperature. Each contact angle reported here is the mean of at least 5 independent measurements [26].

2.4. Antibacterial Analyses

Five hundred μL of bacterial liquid suspensions (2×10^7 cfu/mL) including the Gram-positive bacterium *Staphylococcus aureus* (*S. aureus*) and the Gram-negative bacterium of *Actinobacillus actinomycetemocomitans* (*A. actinomycetemocomitans*) were individually dripped on the surfaces of all the samples. After incubation under a relative humidity of 96% at 37 °C for 4 h in a dark environment, the surfaces of all samples were rinsed three times with phosphate-buffered saline (PBS). The retained bacteria were then fixed with 4% paraformaldehyde (Sigma-Aldrich, St. Louis, MO, USA) for 15 min. Then all the samples were stained with 10 μM SYTO 9, which is a green fluorescent nucleic acid used to stain live and dead bacteria, for 30 min at room temperature. By using an enzyme-linked immunosorbent assay reader (Synergy HT, BioTek Instruments, Winooski, VT, USA), the adhered bacteria on the samples were quantified by measuring the fluorescence at 488 nm.

2.5. Biocompatibility Tests of Cytotoxicity and Cell Viability

To evaluate the biocompatible characteristics of all the samples with specific element compositions, the cytotoxicity test based on ISO 10993-5 assay and cell viability study were performed by using two cell lines, human skin fibroblasts (HSF) and human osteosarcoma (MG-63). HSF (product No. BCRC 60153) and MG-63 (product No. BCRC 60279) cells were purchased from Bioresource Collection and Research Center (Hsinchu City, Taiwan). According to the procedure of cytotoxicity test suggested from the standard of ISO 10993-5, each kind of the samples, such as the the regular culture medium (DMEM), uncoated Ti (control), specimens with Ta_2O_5 and Ta(Zn)O thin films, was individually immersed into serum-free medium with gently shaking at 4 °C. After 72 h, the conditioned medium of each kind of samples was collected by passing through the 0.22 μm sterile filter for the ISO 10993-5 assay. HSF or MG-63 cells cultured in plates with Dulbecco's Modified Eagle Medium (DMEM, Gibco, Carlsbad, CA, USA) were incubated at 37 °C in 5% CO_2 overnight, and then replaced with the conditioned medium above for further 24-h culture. Then a 3-(4,5-dimethylthiazol-2-yl)-2,5-diphenyltetrazolium bromide (MTT assay) was performed. MTT, which is one method to detect the activity of mitochondria in live cells, shows the cell viability. All cells were washed twice by Dulbecco's phosphate-buffered saline (DPBS) and cultured for another 4 h with MTT-contained medium at 37 °C with 5% CO_2. The intracellular purple formazon in live cells was eluted by DMSO (Sigma), quantified by using the ELISA reader (Molecular Devices, San Jose, CA, USA) at the absorbance of 570 nm with SoftMax Pro software (Molecular Devices, San Jose, CA, USA). The measuring procedure was performed and protected from the light.

To examine the effect of all the samples prepared in this study on the cell viability, HSF and MG-63 cells were also cultured directly onto the surface of the uncoated Ti, Ta_2O_5 and Ta(Zn)O films and incubated at 37 °C in 5% CO_2 for 48 h. All cells were then washed twice by Dulbecco's phosphate-buffered saline (DPBS) and cultured for another 4 h with MTT-contained medium at 37 °C with 5% CO_2. The intracellular purple formazon in live cells was eluted by DMSO (Sigma), quantified by using the ELISA reader (Molecular Devices, San Jose, CA, USA) at the absorbance of 570 nm with SoftMax Pro software (Molecular Devices, San Jose, CA, USA). The measuring procedure was performed and protected from the light.

2.6. Statistical Analyses

The ANOVA analysis tool of variance was used to perform the statistical analyses of the contact angle, antibacterial analyses, cytotoxicity and cell viability of all samples. Post hoc pairwise comparisons were conducted precisely using the Turkey method. Each experiment was independently performed and duplicated and the differences were considered significant at the $p < 0.05$ level. SPSS v19 (IBM Corporation, Armonk, NY, USA) was used for statistical analyses.

3. Results and Discussion

3.1. Microstructure and Surface Morphology Analyses

In this study, a field-emission scanning electron microscope and a 3D laser scanning microscope were used to observe the surface morphology of the specimens treated by PEO and HiPIMS, as shown in Figure 2. When a pulsed DC was applied on the uncoated Ti during the PEO treatment, a porous oxide layer was formed, and the surfaces revealed numerous micron-sized pores. The SEM images revealed that the prepared tantalum oxide and Ta(Zn)O films fully cover the porous TiO_2 surface generated by PEO. Followed by the oxide deposition by HiPIMS, the PEO-pretreated samples were still coated with tantalum oxide and Ta(Zn)O, and still retained their porous structure. G. Greczynski et. al. [27] showed that in reactive processes of HiPIMS, metal reactivity, which controlled the rate of compound formation at the target, was expected to be a decisive parameter, as it affected the target sputtering condition and the extent of gas rarefaction. It was found that the film deposited using HiPIMS at low substrate temperature below 80 °C allowed the sputtered particles to gain an optimal energy and form a compact layer on the surface of the substrate [28]. EDS was employed to measure the elemental composition of the oxide films. The chemical composition of the deposited tantalum oxide films comprised 21.7 ± 1.8 at.% of Ta and 78.3 ± 2.1 at.% of O, which implied the formation of stoichiometric tantalum pentoxide (Ta_2O_5) contained a higher fraction of oxygen, and indicated the presence of surface oxides on the surface [26]. The chemical composition of the deposited Ta(Zn)O films comprised 20.8 ± 2.3 at.% of Ta, 4.4 ± 1.3 at.% of Zn, and 74.8 ± 2.7 at.% of O. It showed that the Ta(Zn)O films deposited from Ta and Zn targets had small content of Zn (4.4 ± 1.3 at.%). Philip D. Racka et. al. [29] had revealed that the condition of small fraction of zinc in the films was as a result of the preferential resputtering of Zn versus Ta.

3D Laser microscopy was adopted to examine the surface roughness of the specimens. The surface roughness Ra of the untreated Ti was 0.6 ± 0.15 μm. After the PEO treatment and HiPIMS deposition, the Ta_2O_5 deposited Ti had higher surface roughness (mean roughness Ra = 2.95 ± 0.8 μm; point height of irregularities Rz = 25.3 ± 2.3 μm). The Ta(Zn)O deposited Ti also had higher surface roughness (mean roughness Ra = 3.4 ± 1.1 μm; point height of irregularities Rz = 30.2 ± 2.7 μm). This surface roughening and porous structure may improve in vivo biocompatibility, bone integration, and implant stability [30]. Low-energy metal-ion irradiation of the growing Ta_2O_5 and Ta(Zn)O films during HiPIMS has been shown to provide microstructure densification, without introducing incomplete surface diffusion developed during the deposition of DC magnetron sputtering [31,32]. The results verified that the Ta_2O_5 and Ta(Zn)O coated samples, which had undergone PEO, retained their porous surfaces with high surface roughness. Measured by the Calo test, the coating consisted of ~6.5 μm thick PEO-treated oxide layer and 0.2~0.3 μm thick Ta_2O_5 and Ta(Zn)O coating layer which was prepared by HiPIMS for 15 min.

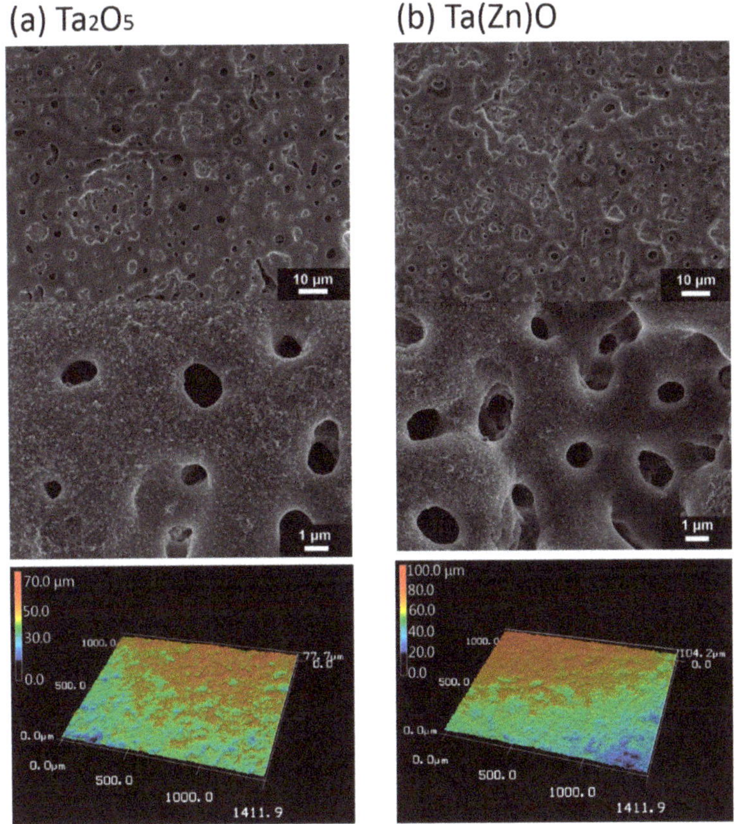

Figure 2. SEM images and 3D surface morphologies of the (**a**) Ta_2O_5 and (**b**) Ta(Zn)O coated titanium samples.

Surface properties of thin films including hydrophilicity and hydrophobicity are explored to improve materials response in biological environments. Surface bio-functionalization routes based on vacuum deposition techniques, together with advances in surface engineering of biomaterials, are important [33]. A hydrophobic surface is likely to lower biocompatibility, whereas a highly hydrophilic surface promotes cell adhesion, migration, and growth. Accordingly, the wettability of the interface between fluid and medical implants should be balanced because the morphology and film composition affect wettability and hydrophobicity. The water contact angles of the untreated Ti, Ta_2O_5, and Ta(Zn)O coated samples were shown in Figure 3. Each point in the figure is an average value of at least ten measurements. The results demonstrated that the pure Ti substrate, Ta_2O_5, and Ta(Zn)O specimen had contact angles of 43.13 ± 10.0°, 3.3 ± 0.36°, and 16.75 ± 3.7°, respectively. It showed obviously that the Ta_2O_5 film on rough and porous Ti was capable of enhancing surface wettability, and adding Zn to thin films increased surface hydrophobicity as compared to Ta_2O_5. Nevertheless, the wettability of Ta(Zn)O was higher than that of the untreated Ti specimen. The hydrophilicity of Ti substantially improved after pretreatment with PEO and coating with Ta_2O_5 and Ta(Zn)O. Surface chemistry and topography separately or together play important roles in the cell response to the oxidized Ti. Surface roughening and hydrophilicity may improve in vivo cell integration and implant stability [34].

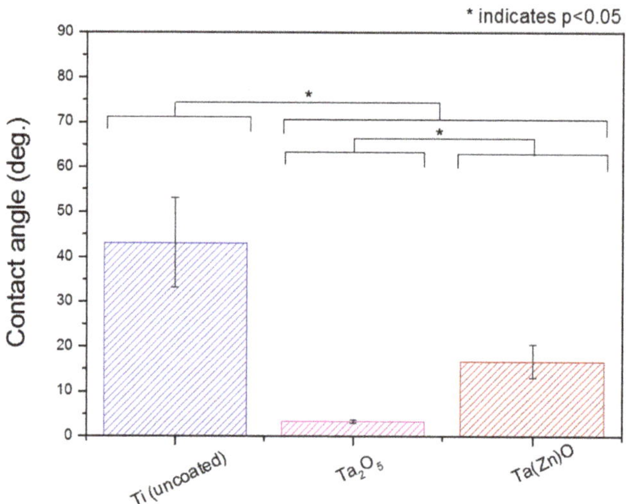

Figure 3. Contact angles of the uncoated Ti, and Ta_2O_5 and TaZnO coated titanium samples. The mean vales of contact angles were different significantly among the untreated Ti, and Ta_2O_5 and Ta(Zn)O coated titanium samples ($p < 0.05$). Post-hoc pairwise comparisons were conducted by the Turkey test. The differences were considered significant when $p < 0.05$.

In order to investigate the oxidation states of elements present in Ta_2O_5 and Ta(Zn)O, XPS was carried out and the results were shown in Figure 4. The chemical bonds of the Ta_2O_5 and Ta(Zn)O coatings were measured and identified. As shown in Figure 4a, the XPS results of Ta_2O_5 coated Ti showed that Ta_2O_5 were deposited on PEO treated Ti, and the shape and position of the XPS peaks (binding energies) clearly indicated the formation of tantalum oxide thin films on Ti [26]. The doublet at 25.7~27.5 eV confirmed the Ta oxidation state (Ta_2O_5) [35,36]. The results of XPS spectra of the Ta(Zn)O samples, as shown in Figure 4b, are similar to the Ta_2O_5 in addition to the presence of Zn2p spectra. The chemical composition of the deposited Ta(Zn)O films measured by XPS consisted of 19.7 ± 1.9 at.% of Ta, 5.3 ± 1.5 at.% of Zn, and 75.0 ± 3.8 at.% of O. The results were similar to EDS. Figure 4b shows the Ta4f, Zn2p and O1s core level XPS spectra of the Ta(Zn)O samples. Zn 2P peaks had the positions located at 1021.7 and 1044.7 eV for the Ta(Zn)O thin films, corresponding to the chemical state of Zn $2p_{3/2}$ and $2p_{1/2}$ in ZnO [37]. The highly intense broad bands of oxygen confirmed the presence of Zn-O and Ta-O bonds. It confirmed the formation of ZnO and Ta_2O_5 composite structure, and verified that Zn was successfully incorporated into Ta_2O_5 to form Ta(Zn)O using HiPIMS.

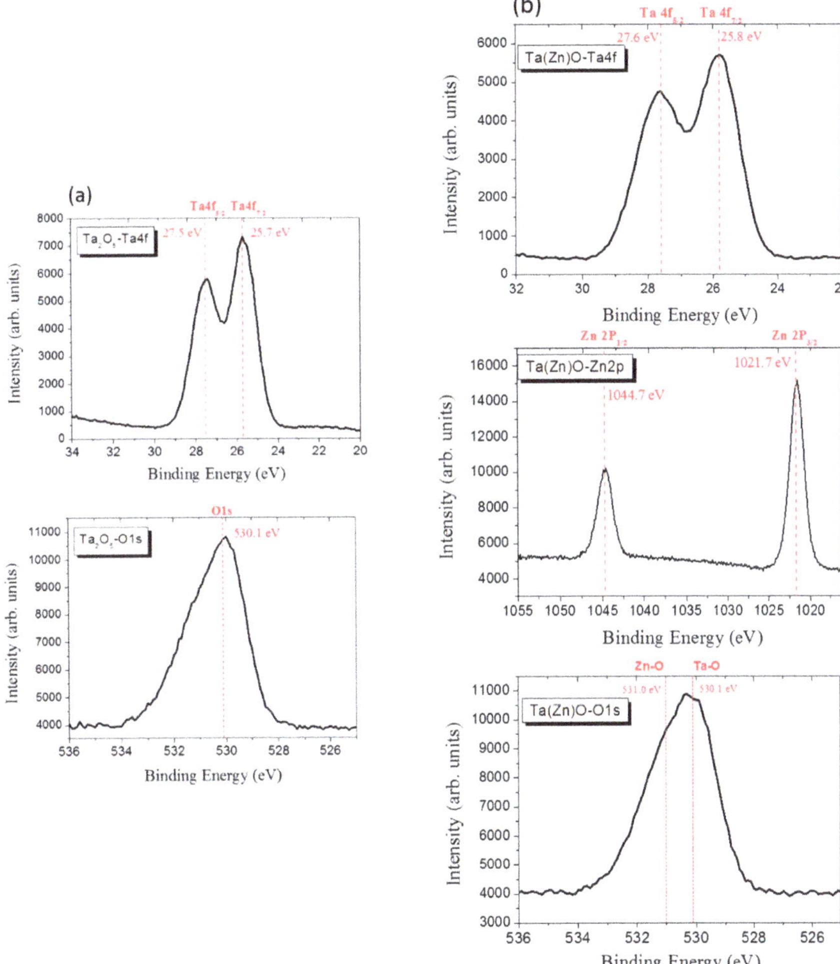

Figure 4. The Ta4f, Zn2p and O1s core level XPS spectra of the (**a**) Ta_2O_5 and (**b**) Ta(Zn)O thin film on Ti.

An X-ray diffractometer was used to analyze the crystallinity of the Ta_2O_5 and Ta(Zn)O thin films. Only the characteristic peaks of the TiO_2 layer were observed for the Ta_2O_5 specimen (Figure 5). It showed that the Ta_2O_5 sample exhibited characteristic peaks of anatase and rutile TiO_2 crystalline phases. The PEO-pretreated Ta_2O_5 sample showed characteristic peaks of TiO_2 but not crystalline Ta_2O_5. The crystalline anatase and rutile TiO_2 were formed by PEO treatment. The Ta(Zn)O films had characteristic peaks of the TiO_2 layer, which was formed by PEO, and the ZnO crystal phases. It indicated that the deposited Ta(Zn)O contained amorphous Ta_2O_5 and crystalline ZnO. The development of the crystalline ZnO structure during growth was controlled by the HiPIMS deposition process [38,39].

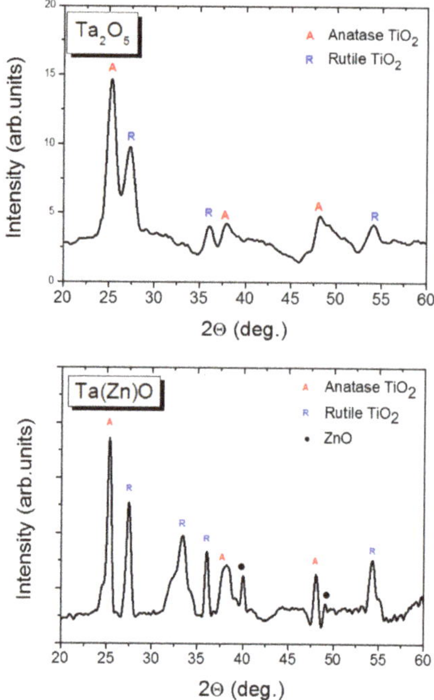

Figure 5. X-Ray diffraction spectra of the Ta_2O_5 and Ta(Zn)O coated Ti with PEO surface pretreatment.

3.2. Antibacterial Properties

SYTO 9 stain was applied to examine the antibacterial property of the untreated Ti, Ta_2O_5, and Ta(Zn)O specimens against *S. aureus* and *A. actinomycetemcomitans*. In Figure 6, the relative fluorescence intensity represents the bacterial residue with STYO 9 stain. Lower relative fluorescence intensity indicates the potential and better capability of the antibacterial ability. The Ta_2O_5 and Ta(Zn)O deposited by HiPIMS showed lower intensity of the relative fluorescence with Syto 9 stain, and the Ta(Zn)O exhibited the lowest intensity in both of the *S. aureus* and *A. actinomycetemcomitans* bacterial colonies (Figure 6). The antibacterial capability of ZnO has been proved in previous studies [22–24]. The content ratio of Zn in the surface coating is a critical issue because the zinc-oxide not only has the ability of improving the antibacterial properties but also increases the cell cytotoxicity [40]. To maintain both of the characteristics of biocompatibility and antibacterial ability, it is important to control the content ratio and manufacture procedure of ZnO-contained coatings.

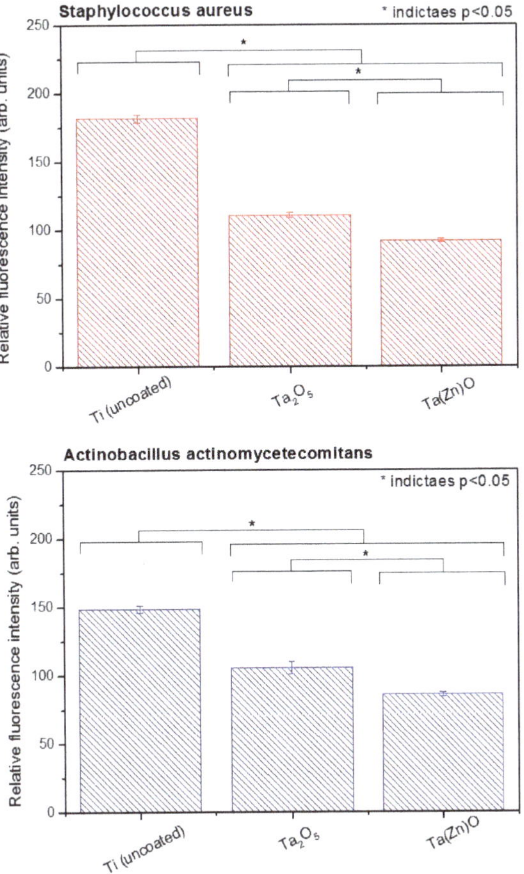

Figure 6. Antibacterial properties of each specimen against S. aureus and A. actinomycetemcomitans. The relative fluorescence intensity represents the bacterial residue with STYO 9 stain, and antibacterial properties among specimens against *S. aureus* and *A. actinomycetemcomitans* were different significantly ($p < 0.05$). Post-hoc pairwise comparisons were conducted by the Turkey test. The differences were considered significant when $p < 0.05$.

3.3. Biocompatibility Tests of Cytotoxicity and Cell Viability

To compare the biocompatible characteristic of the specimens of the uncoated Ti, Ta_2O_5, and Ta(Zn)O with the pretreatment of PEO in the soft tissue- and hard tissue-derived cells, the cytotoxicity and cell viability tests were individually performed with both HSF and MG-63 cells and quantified by MTT assay. Figure 7 showed the results of the cytotoxicity test of all specimens in HSF and MG-63 cells, and Figure 8 showed the results of the cell viability. In Figure 7, HSF cells were cultured with the conditioned medium collected from the untreated Ti, Ta_2O_5, and Ta(Zn)O coatings. Based on the guideline of ISO 10993-5, the group of cells cultured with medium is recommended as the control group for the further comparison with cells cultured with the extraction from any kinds of materials. Besides, the group of the untreated Ti specimen was also essential in this study. Although Ti has been considered as a biomaterial with desirable biocompatibility and mechanical properties especially for the dental and orthopedic implants, coatings with Ta_2O_5 or Ta(Zn)O represented better cell viability and lower cytotoxicity in HSF cells in this study (Figure 7). No significant difference of the cytotoxicity was found between the specimens with Ta_2O_5 and Ta(Zn)O thin films. ZnO did not deteriorate the cytotoxicity of HSF cells on the Ta(Zn)O coated samples.

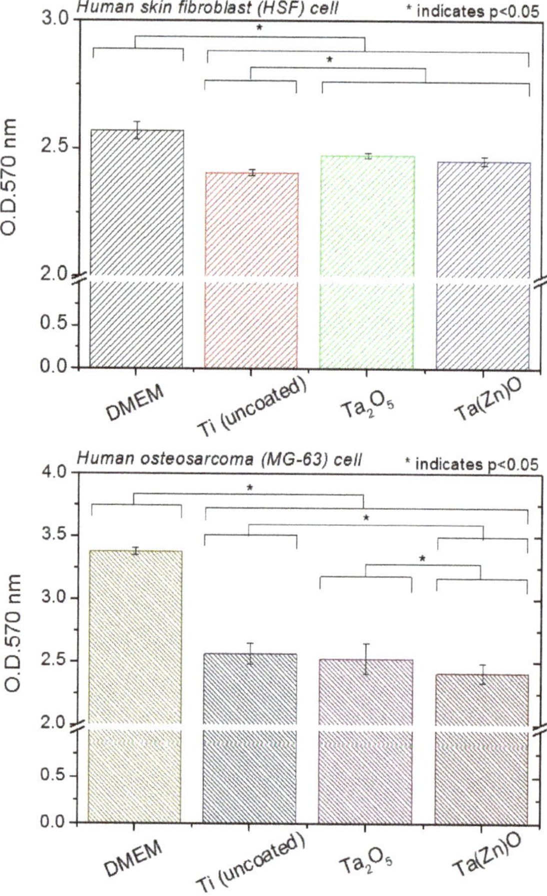

Figure 7. The cytotoxic effect of the uncoated Ti, Ta_2O_5, and Ta(Zn)O films on HSF and MG-63 cells compared to the regular culture medium (DMEM) according to the ISO 10993-5 methodology. The mean vales of O.D. were different significantly among all samples ($p < 0.05$). Post-hoc pairwise comparisons were conducted by the Turkey test and the differences were considered significant when $p < 0.05$.

In Figure 7, the optical density of MG-63 cells cultured with DMEM was much higher than the cells cocultured with the conditioned medium collected from each kind of specimen. No significant difference of the cytotoxicity was observed between Ta_2O_5 and the uncoated Ti specimens. Ta_2O_5 had similar cytotoxicity of MG-63 cells to the pure Ti. Coatings with Ta(Zn)O showed the lowest OD values compared to other groups, indicating that Ta(Zn)O thin films may induce higher level of the cytotoxicity in MG-63 cells. Clinical reports have been provided the evidence that the pure Ti is widely applied to the soft and hard tissues due to its favorable biocompatibility and mechanical properties, especially to the dental and orthopedic implants in clinic [41,42]. Recently, Ta bulk materials and Ta_2O_5 thin films have also been considered as a desirable biomaterial especially applying to clinical implants for the orthopedics and the dentistry [43–46]. Various clinical uses also provide reliable evidences that the addition of Ta of implants improves the hard-tissue growth and healing. Ta is desirable for the adhesion, migration, and proliferation of bone cells. The results above indicated that the deposition of Ta_2O_5 onto the PEO-pretreated Ti substrates with micro porous structure may provide higher affinity to the hard tissue-derived cells for cell growth.

To further study the potential biocompatibility of the uncoated Ti, Ta_2O_5, and Ta(Zn)O films, HSF and MG-63 cells were individually and directly cultured onto each kind of specimen for 48 h followed by MTT assay (Figure 8).

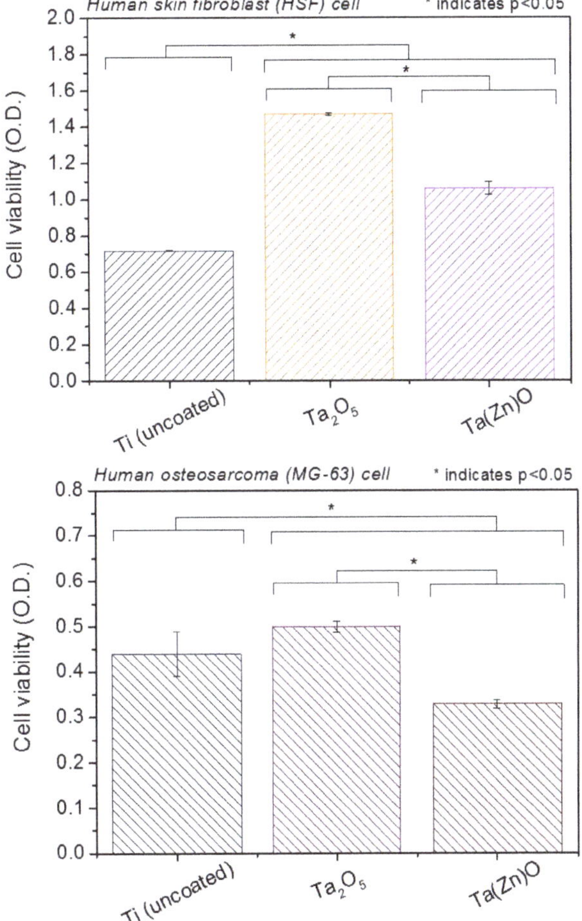

Figure 8. Cell viabilities of fibroblast cells and osteosarcoma cells on each sample. HSF and MG-63 cells were individually and directly cultured onto each kind of specimen for 48 h followed by MTT assay. The mean vales of O.D. were different significantly among all samples ($p < 0.05$). Post-hoc pairwise comparisons were conducted by the Turkey test and the differences were considered significant when $p < 0.05$.

In HSF cells, the results showed that the cell viability of Ta_2O_5 films was not only the greatest compared to the other two kinds of specimens, but also was almost two-fold higher than the uncoated Ti specimen. The thin films with Ta(Zn)O even showed almost 1.5-fold level of the cell viability compared to the uncoated Ti specimen. It indicated that the Ta-contained oxide thin films deposited onto the surface of Ti substrates may enhance the biocompatibility of HSF cells. It is known that cells perceive the surface landscape of Ti implants at very different scales ranging from the molecular level to nanoscale. Vaidulych et al. [47] had studied the convex and concave surface in the nano-scale range of Ta_2O_5 thin films for tailoring the osteoblast adhesion. The surface morphology affected the cell adhesion. Their results showed that primary human osteoblasts were found to sustain the viability on all kinds of these nano-scale convex and concave surfaces; however, Ta_2O_5 with the concave topography showed restrained adhesion of the cells. In this study, the biocompatible effect of Ta_2O_5 and Ta(Zn)O specimens with rough and porous structure in the range of micro-scale was obviously found in HSF cells. The biocompatible effect of Ta_2O_5 specimens in MG-63 cells was consistent with the findings observed in HSF cells. In MG-63 cells, the cell viability of Ta_2O_5 films was higher than the uncoated Ti specimen. Vannozzi, L. et al. [48] studied the ZnO effect on the cell adhesion and viability of composite thin films based on a blend of poly(ethylene glycol)-block-poly(ε-caprolactone) methyl ether (PEG-b-PCL) and poly(l-lactic acid), doped with ZnO nanoparticles. In-vitro tests were carried out with cells of the musculoskeletal apparatus (fibroblasts, osteoblasts, chondrocytes, and myoblasts). All cell types showed good adhesion and viability on the ZnO-doped thin film samples. They found that a higher content of ZnO nanoparticles in the matrix demonstrated higher bioactivity. However, in this study, the coatings with Ta(Zn)O showed the lowest level of the cell viability in MG-63 cells, indicating that the content of Zn may repress the cell viability of MG-63 cells. Safety issues may arise in relation to a possible release of ZnO from the matrix. Zn ions dissociated from ZnO could be associated with cytotoxic phenomena to MG-63 cells. Toxicity was influenced by the ZnO concentration.

These findings above showed that the biological effects of each kind of specimens on the cytotoxicity or the cell viability in the soft tissue-derived cells (HSF cells) was not exactly as same as the results observed in the hard tissue-derived cells (MG-63 cells). For MG-63 cells, the osteoblast-like cells are critical to the surface composition and the microstructure of the materials. It has been proved that the biomaterial surface with the porous structures is better for the osteoblastic cell adhesion, migration, and proliferation [49]. However, the behavior of the fibroblast-like HSF cells is not easily affected by the surface roughness of the materials [50], and in a bacterial environment, the proper surface may give soft tissue-derived cells a better chance to compete with bacteria to cover on the surface of Ti implants [51,52]. This study shows that there is a close relationship between the amount and chemical composition of the thin films serving as a support of the cells.

4. Conclusions

In this study, PEO were mainly applied to form a well-distributed porous TiO_2 surface, and HiPIMS was further adopted to deposit Ta_2O_5 and Ta(Zn)O coatings on the porous surface of Ti to discuss the material properties, antibacterial performance, and biocompatibility of the thin films.

The conclusions are shown as follows:

1) Scanning electron microscopy and 3D laser microscopy were used to observe the morphology of the specimens. The surfaces of the Ti specimens that had been subjected to PEO were fully covered with a TiO_2 layer. Through the use of HiPIMS, the deposition of the Ta_2O_5 and Ta(Zn)O films would not affect the porous structure on the specimen surface. The Ta_2O_5 and Ta(Zn)O-deposited Ti had high surface roughness and this surface roughening and porous structure may influence the biocompatibility. The results of water contact angle measurement demonstrated that the pure Ti substrate, Ta_2O_5, and Ta(Zn)O specimen had contact angles of $43.13 \pm 10.0°$, $3.3 \pm 0.36°$, and $16.75 \pm 3.7°$, respectively. The deposited Ta_2O_5 coating on the porous Ti was capable of enhancing surface wettability, and adding Zn to thin films increased surface hydrophobicity as compared to Ta_2O_5. The deposited Ta(Zn)O contained amorphous Ta_2O_5 and crystalline ZnO. The development of the crystalline ZnO structure during growth is controlled by the HiPIMS deposition process.
2) Antibacterial properties against Gram-positive and Gram-negative bacteria were performed. Compared to the untreated Ti, both specimens with Ta_2O_5 and Ta(Zn)O thin films showed lower intensity of the relative fluorescence with Syto 9 stain in both these two bacterial colonies,

and they had improved antibacterial abilities against *S. aureus* and *A. actinomycetemcomitans*. The Ta(Zn)O deposited by HiPIMS exhibited the lowest intensity in both of the *S. aureus* and *A. actinomycetemcomitans* bacterial colonies, and it showed the best antibacterial performance.

3) The results of the ISO-10993-5 cell cytotoxicity and cell viability MTT assay tests revealed that the Ta_2O_5 or Ta(Zn)O coated Ti had better cell viability and lower cytotoxicity in HSF cells, and both coatings possessed high biocompatibilities in this study. The cell activity of Ta(Zn)O coated Ti decreased slightly when reacting to the MG63 cells. The result showed that the Ta_2O_5 coating on Ti surface with porous structure improved adhesion, migration, and proliferation for the osteoblastic cell. Due to the different properties of cells, HSF and MG-63 cells represented different behaviors of cell cytoxicity and cell viability in each kind of coating specimens on Ti.

Author Contributions: Conceptualization: Y.-Y.C., H.-L.H.; methodology: Y.-Y.C., H.-L.H.; validation: Y.-Y.C.; formal analysis: Y.-Y.C., H.-L.H., M.-T.T.; investigation: Y.-Y.C.; resources: Y.-Y.C.; data Ccration: Y.-Y.C., H.-L.H.; writing—original draft preparation: Y.-Y.C., Y.-J.L.; writing—review and editing: Y.-Y.C., H.-L.H., M.-T.T.; visualization: J.-T.H.; supervision: Y.-Y.C.; project administration: Y.-Y.C., H.-L.H.; funding acquisition: Y.-Y.C., J.-T.H. All authors have read and agreed to the published version of the manuscript.

Funding: This research was supported by the Ministry of Science and Technology (MOST 107-2218-E-131 -001 and MOST 108-2221-E-150-020-MY3) of Taiwan. This work was also financially supported by the "High Entropy Materials Center" from The Featured Areas Research Center Program within the framework of the Higher Education Sprout Project by the Ministry of Education (MOE) and from the Project MOST 108-3017-F-007-002- by Ministry of Science and Technology (MOST) in Taiwan.

Acknowledgments: The instrumental assistance from the Common Lab. for Micro/Nano Sci. and Tech. of National Formosa University is sincerely appreciated.

Conflicts of Interest: The authors declare no conflict of interest.

References

1. Xiao, M.; Chen, Y.; Biao, M.; Zhang, X.; Yang, B. Bio-functionalization of biomedical metals. *Mater. Sci. Eng. C* **2017**, *70*, 1057–1070. [CrossRef] [PubMed]
2. Chen, Q.; Thouas, G.A. Metallic implant biomaterials. *Mater. Sci. Eng. R* **2015**, *87*, 1–57. [CrossRef]
3. Devgan, S.; Sidhu, S.S. Evolution of surface modification trends in bone related biomaterials: A review. *Mater. Chem. Phys.* **2019**, *233*, 68–78. [CrossRef]
4. Chouirfa, H.; Bouloussa, H.; Migonney, V.; Falentin-Daudré, C. Review of titanium surface modification techniques and coatings for antibacterial applications. *Acta Biomater.* **2019**, *83*, 37–54. [CrossRef]
5. Villafuerte, K.R.V.; Martinez, C.D.J.H.; Dantas, F.T.; Carrara, H.H.A.; dos Reis, F.J.C.; Palioto, D.B. The impact of chemotherapeutic treatment on the oral microbiota of patients with cancer: A systematic review. *Oral Surg. Oral Med. Oral Pathol. Oral Radiol.* **2018**, *125*, 552–566. [CrossRef]
6. Mortazavi, G.; Jiang, J.; Meletis, E.I. Investigation of the plasma electrolytic oxidation mechanism of titanium. *Appl. Surf. Sci.* **2019**, *488*, 370–382. [CrossRef]
7. Tanase, C.E.; Golozar, M.; Best, S.M.; Brooks, R.A. Cell response to plasma electrolytic oxidation surface-modified low-modulus β-type titanium alloys. *Colloids Surf. B* **2019**, *176*, 176–184. [CrossRef]
8. Janson, O.; Gururaj, S.; Pujari-Palmer, S.; Karlsson Ott, M.; Strømme, M.; Engqvist, H.; Welch, K. Titanium surface modification to enhance antibacterial and bioactive properties while retaining biocompatibility. *Mater. Sci. Eng. C* **2019**, *96*, 272–279. [CrossRef]
9. Tsai, M.-T.; Chang, Y.-Y.; Huang, H.-L.; Hsu, J.-T.; Chen, Y.-C.; Wu, A.Y.-J. Characterization and antibacterial performance of bioactive Ti–Zn–O coatings deposited on titanium implants. *Thin Solid Films* **2013**, *528*, 143–150. [CrossRef]
10. Wang, Q.; Qiao, Y.; Cheng, M.; Jiang, G.; He, G.; Chen, Y.; Zhang, X.; Liu, X. Tantalum implanted entangled porous titanium promotes surface osseointegration and bone ingrowth. *Sci. Rep.* **2016**, *6*, 26248. [CrossRef]
11. Li, Y.; Zhao, T.; Wei, S.; Xiang, Y.; Chen, H. Effect of Ta2O5/TiO2 thin film on mechanical properties, corrosion and cell behavior of the NiTi alloy implanted with tantalum. *Mater. Sci. Eng. C* **2010**, *30*, 1227–1235. [CrossRef]
12. Levine, B.R.; Sporer, S.; Poggie, R.A.; Della Valle, C.J.; Jacobs, J.J. Experimental and clinical performance of porous tantalum in orthopedic surgery. *Biomaterials* **2006**, *27*, 4671–4681. [CrossRef] [PubMed]

13. Hanzlik, J.A.; Day, J.S.; Contributors, A.; Group, I.R.S. Bone ingrowth in well-fixed retrieved porous tantalum implants. *J. Arthroplast.* **2013**, *28*, 922–927. [CrossRef] [PubMed]
14. Issack, P.S. Use of porous tantalum for acetabular reconstruction in revision hip arthroplasty. *JBJS* **2013**, *95*, 1981–1987. [CrossRef]
15. Bencharit, S.; Byrd, W.C.; Altarawneh, S.; Hosseini, B.; Leong, A.; Reside, G.; Morelli, T.; Offenbacher, S. Development and applications of porous tantalum trabecular metal-enhanced titanium dental implants. *Clin. Implant Dent. Relat. Res.* **2014**, *16*, 817–826. [CrossRef]
16. Chang, Y.-Y.; Huang, H.-L.; Chen, H.-J.; Lai, C.-H.; Wen, C.-Y. Antibacterial properties and cytocompatibility of tantalum oxide coatings. *Surf. Coat. Technol.* **2014**, *259*, 193–198. [CrossRef]
17. Mareci, D.; Chelariu, R.; Gordin, D.-M.; Ungureanu, G.; Gloriant, T. Comparative corrosion study of Ti–Ta alloys for dental applications. *Acta Biomater.* **2009**, *5*, 3625–3639. [CrossRef]
18. Zhang, Y.; Nayak, T.R.; Hong, H.; Cai, W. Biomedical applications of zinc oxide nanomaterials. *Curr. Mol. Med.* **2013**, *13*, 1633–1645. [CrossRef]
19. Seo, H.-J.; Cho, Y.-E.; Kim, T.; Shin, H.-I.; Kwun, I.-S. Zinc may increase bone formation through stimulating cell proliferation, alkaline phosphatase activity and collagen synthesis in osteoblastic MC3T3-E1 cells. *Nutr. Res. Pract.* **2010**, *4*, 356–361. [CrossRef]
20. Yamaguchi, M.; Yamaguchi, R. Action of zinc on bone metabolism in rats: Increases in alkaline phosphatase activity and DNA content. *Biochem. Pharmacol.* **1986**, *35*, 773–777. [CrossRef]
21. Jin, G.; Cao, H.; Qiao, Y.; Meng, F.; Zhu, H.; Liu, X. Osteogenic activity and antibacterial effect of zinc ion implanted titanium. *Colloids Surf. B* **2014**, *117*, 158–165. [CrossRef] [PubMed]
22. Raghupathi, K.R.; Koodali, R.T.; Manna, A.C. Size-dependent bacterial growth inhibition and mechanism of antibacterial activity of zinc oxide nanoparticles. *Langmuir* **2011**, *27*, 4020–4028. [CrossRef] [PubMed]
23. Qi, K.; Cheng, B.; Yu, J.; Ho, W. Review on the improvement of the photocatalytic and antibacterial activities of ZnO. *J. Alloys Compd.* **2017**, *727*, 792–820. [CrossRef]
24. Banoee, M.; Seif, S.; Nazari, Z.E.; Jafari-Fesharaki, P.; Shahverdi, H.R.; Moballegh, A.; Moghaddam, K.M.; Shahverdi, A.R. ZnO nanoparticles enhanced antibacterial activity of ciprofloxacin against Staphylococcus aureus and Escherichia coli. *J. Biomed. Mater. Res. Part B* **2010**, *93*, 557–561. [CrossRef] [PubMed]
25. Zhang, X.; Wu, H.; Geng, Z.; Huang, X.; Hang, R.; Ma, Y.; Yao, X.; Tang, B. Microstructure and cytotoxicity evaluation of duplex-treated silver-containing antibacterial TiO2 coatings. *Mater. Sci. Eng. C* **2014**, *45*, 402–410. [CrossRef]
26. Huang, H.-L.; Tsai, M.-T.; Lin, Y.-J.; Chang, Y.-Y. Antibacterial and biological characteristics of tantalum oxide coated titanium pretreated by plasma electrolytic oxidation. *Thin Solid Films* **2019**, *688*, 137268. [CrossRef]
27. Greczynski, G.; Zhirkov, I.; Petrov, I.; Greene, J.E.; Rosen, J. Control of the metal/gas ion ratio incident at the substrate plane during high-power impulse magnetron sputtering of transition metals in Ar. *Thin Solid Films* **2017**, *642*, 36–40. [CrossRef]
28. Alias, R.; Mahmoodian, R.; Genasan, K.; Vellasamy, K.M.; Hamdi Abd Shukor, M.; Kamarul, T. Mechanical, antibacterial, and biocompatibility mechanism of PVD grown silver–tantalum-oxide-based nanostructured thin film on stainless steel 316L for surgical applications. *Mater. Sci. Eng. C* **2020**, *107*, 110304. [CrossRef]
29. Rack, P.D.; Potter, M.D.; Woodard, A.; Kurinec, S. Negative ion resputtering in Ta2Zn3O8 thin films. *J. Vac. Sci. Technol. A* **1999**, *17*, 2805–2810. [CrossRef]
30. Kim, G.; Hong, L.Y.; Jung, J.; Kim, D.-P.; Kim, H.; Kim, I.J.; Kim, J.R.; Ree, M. The biocompatibility of mesoporous inorganic–organic hybrid resin films with ionic and hydrophilic characteristics. *Biomaterials* **2010**, *31*, 2517–2525. [CrossRef]
31. Petrov, I.; Barna, P.B.; Hultman, L.; Greene, J.E. Microstructural evolution during film growth. *J. Vac. Sci. Technol. A* **2003**, *21*, S117–S128. [CrossRef]
32. Mirica, E.; Kowach, G.; Du, H. Modified Structure Zone Model to Describe the Morphological Evolution of ZnO Thin Films Deposited by Reactive Sputtering. *Cryst. Growth Des.* **2004**, *4*, 157–159. [CrossRef]
33. Mozetic, M.; Vesel, A.; Primc, G.; Eisenmenger-Sittner, C.; Bauer, J.; Eder, A.; Schmid, G.H.S.; Ruzic, D.N.; Ahmed, Z.; Barker, D.; et al. Recent developments in surface science and engineering, thin films, nanoscience, biomaterials, plasma science, and vacuum technology. *Thin Solid Films* **2018**, *660*, 120–160. [CrossRef]
34. Sul, Y.-T. The significance of the surface properties of oxidized titanium to the bone response: Special emphasis on potential biochemical bonding of oxidized titanium implant. *Biomaterials* **2003**, *24*, 3893–3907. [CrossRef]

35. Chennakesavulu, K.; Reddy, M.M.; Reddy, G.R.; Rabel, A.M.; Brijitta, J.; Vinita, V.; Sasipraba, T.; Sreeramulu, J. Synthesis, characterization and photo catalytic studies of the composites by tantalum oxide and zinc oxide nanorods. *J. Mol. Struct.* **2015**, *1091*, 49–56. [CrossRef]
36. Li, J.; Dai, W.; Wu, G.; Guan, N.; Li, L. Fabrication of Ta2O5 films on tantalum substrate for efficient photocatalysis. *Catal. Commun.* **2015**, *65*, 24–29. [CrossRef]
37. NuLi, Y.-N.; Fu, Z.-W.; Chu, Y.-Q.; Qin, Q.-Z. Electrochemical and electrochromic characteristics of Ta2O5–ZnO composite films. *Solid State Ionics* **2003**, *160*, 197–207. [CrossRef]
38. Rezek, J.; Novák, P.; Houška, J.; Pajdarová, A.D.; Kozák, T. High-rate reactive high-power impulse magnetron sputtering of transparent conductive Al-doped ZnO thin films prepared at ambient temperature. *Thin Solid Films* **2019**, *679*, 35–41. [CrossRef]
39. Reed, A.N.; Shamberger, P.J.; Hu, J.J.; Muratore, C.; Bultman, J.E.; Voevodin, A.A. Microstructure of ZnO thin films deposited by high power impulse magnetron sputtering. *Thin Solid Films* **2015**, *579*, 30–37. [CrossRef]
40. Chang, Y.-Y.; Lai, C.-H.; Hsu, J.-T.; Tang, C.-H.; Liao, W.-C.; Huang, H.-L. Antibacterial properties and human gingival fibroblast cell compatibility of TiO 2/Ag compound coatings and ZnO films on titanium-based material. *Clin. Oral Investig.* **2012**, *16*, 95–100. [CrossRef]
41. Geetha, M.; Singh, A.; Asokamani, R.; Gogia, A. Ti based biomaterials, the ultimate choice for orthopaedic implants—A review. *Prog. Mater. Sci.* **2009**, *54*, 397–425. [CrossRef]
42. Sidambe, A. Biocompatibility of advanced manufactured titanium implants—A review. *Materials* **2014**, *7*, 8168–8188. [CrossRef] [PubMed]
43. Meng, F.; Li, Z.; Liu, X. Synthesis of tantalum thin films on titanium by plasma immersion ion implantation and deposition. *Surf. Coat. Technol.* **2013**, *229*, 205–209. [CrossRef]
44. Mei, S.; Yang, L.; Pan, Y.; Wang, D.; Wang, X.; Tang, T.; Wei, J. Influences of tantalum pentoxide and surface coarsening on surface roughness, hydrophilicity, surface energy, protein adsorption and cell responses to PEEK based biocomposite. *Colloids Surf. B* **2019**, *174*, 207–215. [CrossRef]
45. Sagherian, B.H.; Claridge, R.J. The Use of Tantalum Metal in Foot and Ankle Surgery. *Orthop. Clin. N. Am.* **2019**, *50*, 119–129. [CrossRef]
46. De Paolis, M.; Zucchini, R.; Romagnoli, C.; Romantini, M.; Mariotti, F.; Donati, D.M. Middle term results of tantalum acetabular cups in total hip arthroplasty following pelvic irradiation. *Acta Orthop. Traumatol. Turc.* **2019**, *53*, 165–169. [CrossRef]
47. Vaidulych, M.; Pleskunov, P.; Kratochvíl, J.; Mašková, H.; Kočová, P.; Nikitin, D.; Hanuša, J.; Kyliána, K.; Štěrbabc, J.; Biederman, H.; et al. Convex vs concave surface nano-curvature of Ta_2O_5 thin films for tailoring the osteoblast adhesion. *Surf. Coat. Technol.* **2020**, *393*, 125805–125812. [CrossRef]
48. Vannozzi, L.; Gouveia, P.J.; Pingue, P.; Canale, C.; Ricotti, L. Novel ultra-thin films based on a blend of PEG-b-PCL and PLLA and doped with ZnO nanoparticles. *ACS Appl. Mater. Interfaces* **2020**, *12*, 21398–21410. [CrossRef]
49. Martin, J.; Schwartz, Z.; Hummert, T.; Schraub, D.; Simpson, J.; Lankford, J., Jr.; Dean, D.D.; Cochran, D.L.; Boyan, B. Effect of titanium surface roughness on proliferation, differentiation, and protein synthesis of human osteoblast-like cells (MG63). *J. Biomed. Mater. Res.* **1995**, *29*, 389–401. [CrossRef]
50. Richards, R. The effect of surface roughness on fibroblast adhesion in vitro. *Injury* **1996**, *27*, S/C38–S/C43. [CrossRef]
51. Zhao, B.; Van Der Mei, H.C.; Subbiahdoss, G.; de Vries, J.; Rustema-Abbing, M.; Kuijer, R.; Busscher, H.J.; Ren, Y. Soft tissue integration versus early biofilm formation on different dental implant materials. *Dent. Mater.* **2014**, *30*, 716–727. [CrossRef] [PubMed]
52. Wytrwal, M.; Koczurkiewicz, P.; Zrubek, K.; Niemiec, W.; Michalik, M.; Kozik, B.; Szneler, E.; Bernasik, A.; Madeja, Z.; Nowakowska, M.; et al. Growth and motility of human skin fibroblasts on multilayer strong polyelectrolyte films. *J. Colloid Interface Sci.* **2016**, *461*, 305–316. [CrossRef] [PubMed]

Article

Microstructure and Mechanical Properties of Titanium–Equine Bone Biocomposites

Wonki Jeong [1], Se-Eun Shin [1,*] and Hyunjoo Choi [2]

1 Department of Materials Science and Metallurgical Engineering, Sunchon National University, Suncheon 57922, Korea; show4266@naver.com
2 School of Advanced Materials Engineering, Kookmin University, Seoul 02707, Korea; hyunjoo@kookmin.ac.kr
* Correspondence: shinsen@scnu.ac.kr

Received: 31 March 2020; Accepted: 24 April 2020; Published: 29 April 2020

Abstract: Microstructure and mechanical properties of Ti-6Al-4V/equine bone (EB) composites fabricated by ball milling and spark plasma sintering (SPS) have been investigated. Ti-6Al-4V/EB composites were successfully fabricated by a planetary ball-milling of spherical Ti6Al4V powder and natural EB powder and SPS at 1000 °C within 15 min under 50 MPa. EB was uniformly dispersed in the Ti6Al4V matrix owing to ball-milling, and beta phase transformation temperature of 1000 °C provided phase stability. The composites containing 0.5 wt.% EB exhibit Vickers hardness and elastic modulus of 540.6 HV and 130.5 GPa, respectively. The microstructures and mechanical properties of the composites were observed using scanning electron micrograph and nanoindentation.

Keywords: metal–matrix composites; titanium alloy design; microstructures; mechanical properties; biocomposites; powder metallurgy

1. Introduction

Recently, there has been a high demand for implants for bone dysfunction caused by damages, diseases, and fractured bones by aging or accident, so numerous studies related to biomedical implant materials which can function on the damages or fractured bones by aging or accident have been actively conducted [1–3]. According to the distinct required properties for biomedical implants, limited metals or ceramics were selected. The biomedical implants must not react with a tissue of the human body because it causes biological instability due to corrosion or degradation of the implants [4]. Therefore, not only high strength and low stiffness with lightweight but also corrosion resistance in vivo should be satisfied to apply in biomedical applications. For metals, titanium (Ti) and its alloys, stainless steel, and cobalt–chromium alloys have been extensively reported [5,6].

Among the metals, Ti and its alloys are widely used as implant materials owing to excellent biocompatibility with tissues in the human body and high corrosion resistance and mechanical properties [4,7–9]. The Ti-6 aluminum-4 Vanadium (Ti6Al4V) alloy, the most extensively used Ti alloy, has been used successfully for dental and orthopedic implants, due to its high corrosion resistance and osseointegration by a thin oxide layer formed on the surface in a very short time [10]. The oxide layer exists on the surface of thickness in ~2 nm, which prevents elution of the metal ions and provides the corrosion resistance to withstand the intra-vital corrosion environment [11].

Hence, most biomaterials have used hydroxyapatite ($Ca_{10}(PO_4)_6(OH)_2$) or calcium pyrophosphate ($Ca_2P_2O{\cdot}2H_2O$) coated with Ti6Al4V alloys using several coating methods, such as plasma spraying, thermal spraying, and the hot dipping method to increase the osseointegration with the bones [12–15]. In order to give high osteoconductive property to Ti6Al4V alloy, calcium phosphate-based ceramics such as hydroxyapatite are also coated by plasma spraying [16].

Especially hydroxyapatite has high biocompatibility, bioactivity, osteoconductive property, and osteoinduction, since a lot of researchers have released results of biomaterials containing hydroxyapatite [13,15,17–22]. Meanwhile, biomaterials containing calcium phosphate ($Ca_3(PO_4)_2$) from biological wastes such as pig bones, fish bones, and horse bones have been used as raw materials for mass production of natural hydroxyapatite at low cost, according to recent reports [23–25]. Production of the hydroxyapatite from abundant natural sources rather than synthetic sources is much more economical and eco-friendly. In particular, horse bones are free from foot-and-mouth disease and have a positive effect on the extraction of natural calcium phosphate due to the high supply rate and low price due to the increase in the number of slaughter horses.

However, in the case of metal coated with the synthetically-produced hydroxyapatite, detachment of hydroxyapatite from a metallic substrate occurs, which is a fatal disadvantage [26–28]. For instance, biomedical implants coated with calcium phosphates from the natural bones using the spray method may cause necrosis of cells because, during long-term use, the differences in thermal expansion coefficients and mechanical properties between the calcium phosphate and the metallic materials may cause peeling to occur. Therefore, a study on the Ti6Al4V matrix biocomposites containing hydroxyapatite produced by powder metallurgy has suggested that peeling problems can be avoided and presented high strength and low elastic modulus compared to microcrystalline Ti [29].

The aim of the present study is to suggest the fabrication process of the Ti6Al4V/equine bone (EB) composite using powder metallurgy and overcome the peeling of the hydroxyapatite from the Ti6Al4V. Hence, Ti6Al4V/EB composite powders are produced using a low-energy ball-milling and the composite powders rapidly sintered by spark plasma sintering (SPS) have inhibited grain growth and decomposition of hydroxyapatite. The effect of the EB on the microstructure and mechanical properties has been evaluated as a function of EB weight fraction; a quantitative and qualitative analysis of the composites is also done.

2. Experimental Procedure

2.1. Materials and Methods

Ti6Al4V-EB composites were fabricated by a powder metallurgy route. Commercial powder of pure Ti6Al4V (AP&C, Quebec, Canada) with a mean particle size of 35 μm and a composition of 6.2% Al, 4.1% V, and 0.1% O in a Ti matrix (weight%), and EB powders (Jeju, Korea) with an average powder size of around 1 μm and a composition of 39.4% O, 41.4% Ca, 16.7% P, and 0.56% Mg were used. The Ti6Al4V-EB composite powders of 50 g were ball-milled and utilized by a planetary mill (Pulverisette 4, Fritsch, Idar-Oberstein, Germany) at a rotational speed of 200 RPM for 12 h, followed by a 40 min pause after every 20 min of milling to avoid overheating under air atmosphere. A stainless steel bowl (500 ml) was charged with the Ti6Al4V and EB powders and stainless steel balls without any process control agent. The diameter of the stainless steel balls was 15 mm, and the ball-to-powder weight ratio employed was 5:1. Owing to the shearing mode, EB powders were gradually embedded into the Ti6Al4V powder. The Ti6Al4V and ball-milled Ti6Al4V-EB composite powders were rapidly consolidated using SPS. The composite powders were poured into a 30 mm graphite die with 10 g, and the heating rate was 100 °C/min up to 1000 °C/min, and maintained for 15 min at an applied pressure of 50 MPa under a high vacuum atmosphere. After SPS, the pressure was removed, and the sintered specimen was cooled down to room temperature. SPSed Ti6Al4V and Ti6Al4V-EB composites have dimensions of 30 mm in diameter and 10 mm in height.

2.2. Characterizations

The morphologies of the starting powders, composite powders, and sintered composites were observed using field-emission scanning electron microscopy (FE-SEM, JSM 7001F, JEOL, Tokyo, Japan), and energy dispersive spectroscopy (EDS) revealed the chemical compositions of the powders and the composites. X-ray diffraction (XRD, CN2301, Rigaku, Tokyo, Japan) with a Cu-Kα radiation source (λ = 1.5405 Å) using a step size of 0.02° (2θ), the scanning rate of 1°/min from 20 to 100° to identify the phase constitutions of the composite powders and sintered composites.

To measure elastic modulus of the composites, SPSed specimens with a 4 mm cube were ground flat and mechanically polished up to 4000 grit. An ultrasonic technique was used to determine elastic modulus of the composites (5055PR Pulse receiver 5055PR and Oscilloscope 9354CM oscilloscope, LeCoroy Co., NY, USA). A frequency of 5 MHz was applied; the longitudinal velocity and transverse velocities y within the specimens were determined. Nanoindentation tests were performed on the specimens using a commercial nanohardness tester (Nanoindenter XP, MTS, MN, USA) equipped with a Berkovich indenter and measured more than 10 times per each specimen. In each test, the indenter was driven into the sample surface (loading half-cycle) at a rate of 10 nm/s and the peak load ranges from 0 to 35 mN. The Vickers hardness of the specimens was measured using a micro-Vickers hardness testing machine (Mitutoyo, HM200, Kawasaki, Japan) with an indenter load of 300 N.

3. Results and Discussion

3.1. Microstructures

Figure 1 shows SEM images of starting powders of Ti6Al4V and EB, and ball-milled powders of Ti6Al4V-EB composite powders as a function of the EB weight fraction (wt.%); 0.05, 0.5, and 5 wt.%. The composites containing EB of 0.05, 0.5, and 5 wt.% were expressed as T-0.05EB, T-0.5EB, and T-5EB, respectively. According to the images, the shape of the ball-milled powders was insignificantly changed, but a rough surface iwa shown due to the ball-milling process. During the ball-milling process, physical changes were induced by the impact between the balls and the powders accompanying shear force. Therefore, the size of the powders was slightly increased from 32 to 45 μm, because certain cold welding could occur between the balls [30]. In the process, nanoscale EB powders were attached to the Ti6Al4V powder surface and then gradually embedded in the Ti6Al4V matrix. Accordingly, EB powders were not observed on the Ti6Al4V powder surface in all composite powders as shown in Figure 1d,f,h. From the elemental analysis using EDS, the key chemical composition was maintained as shown in Figure 1i. The main chemical composition of the hydroxyapatite, such as Ca, P, and O, increased as the EB contents increased.

Figure 2 presents the SEM images and EDS mapping of SPSed Ti6Al4V and the Ti6Al4V-EB composites with the weight fractions of EB 0.05, 0.5, and 5 wt.%. As shown in the images, α-Ti with the hexagonal closed packed (HCP) and β-Ti with the body-centered cubic (BCC) phases are clearly shown as dark grey and bright grey, respectively, and a volume fraction of β-Ti phase (f_{β}) is displayed in each of the SEM images. The chemical compositions for each phase from EDS analysis are shown in Table 1. As a reference, defects such as pores and voids were not significantly found in the Ti6Al4V sintered specimen, and the presence and amount of β-Ti were 12.6%. The sintered composites having a porosity below 1% and relative density above 99% on average, excluding Ti6Al4V-5EB composites. Ti6Al4V-5EB composites showed a relative density of 95%, which affects elastic modulus and the hardness results.

Table 1. EDS analysis of the samples with the phase fraction of α-Ti and β-Ti phase from the SEM images of Figure 2.

Element	(a) Ti6Al4V		(b) T-0.05EB		(c) T-0.5EB		(d) T-5EB	
	α-Phase	β-Phase	α-Phase	β-Phase	α-Phase	β-Phase	α-Phase	β-Phase
Al	5.94	5.21	6.90	5.93	6.58	5.55	6.40	6.65
P	-	-	0.04	0.05	0.00	0.09	0.03	0.05
Ca	-	-	0.00	0.03	0.30	0.05	0.18	0.19
O	0.18	0.15	1.29	1.51	1.97	2.54	2.3	4.2
Ti	90.12	89.37	88.69	84.26	88.61	83.54	88.08	79.35
V	3.76	5.28	3.08	8.22	2.54	8.23	3.01	5.37
Totals	100.00	100.00	100.0	100.0	100.0	100.0	100.0	100.0

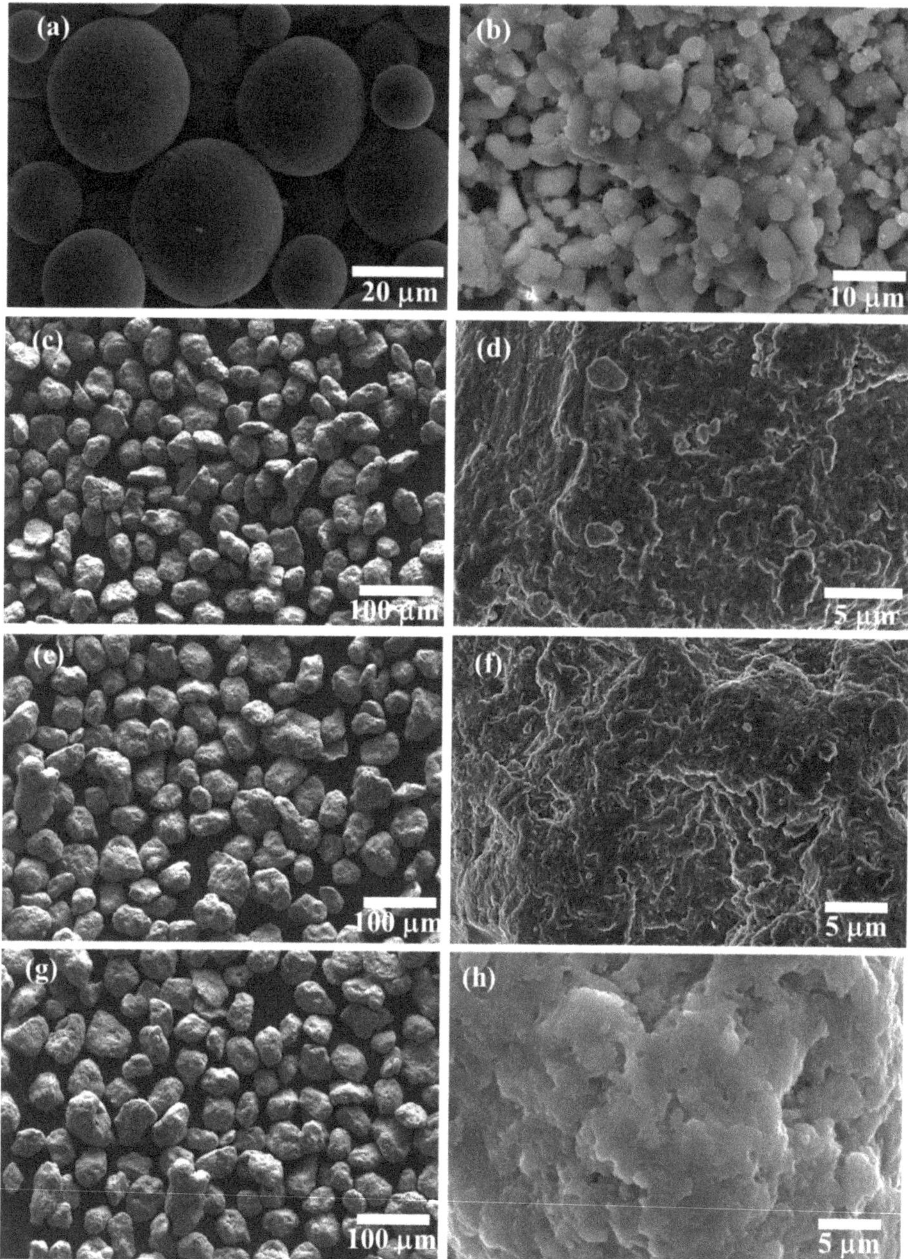

Figure 1. *Cont.*

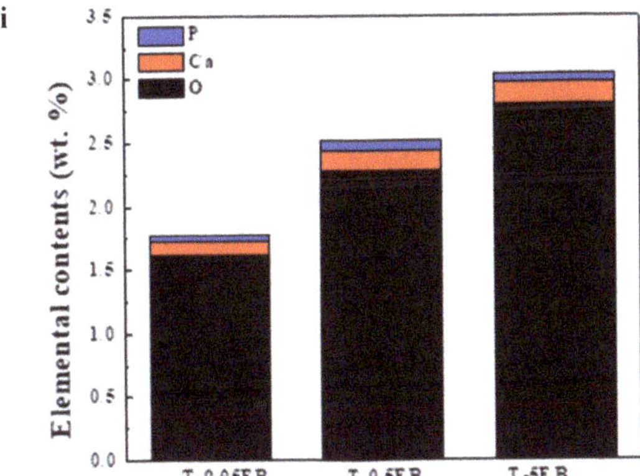

Figure 1. SEM images of starting materials and composite powders as a function of their equine bone (EB) contents; (**a**) Ti6Al4V; (**b**) EB; (**c**,**d**) T-0.05EB; (**e**,**f**) T-0.5EB; and (**g**,**h**) T-5EB. (**i**) Average chemical composition of Ca, P, and O in (**d**), (**f**), and (**h**).

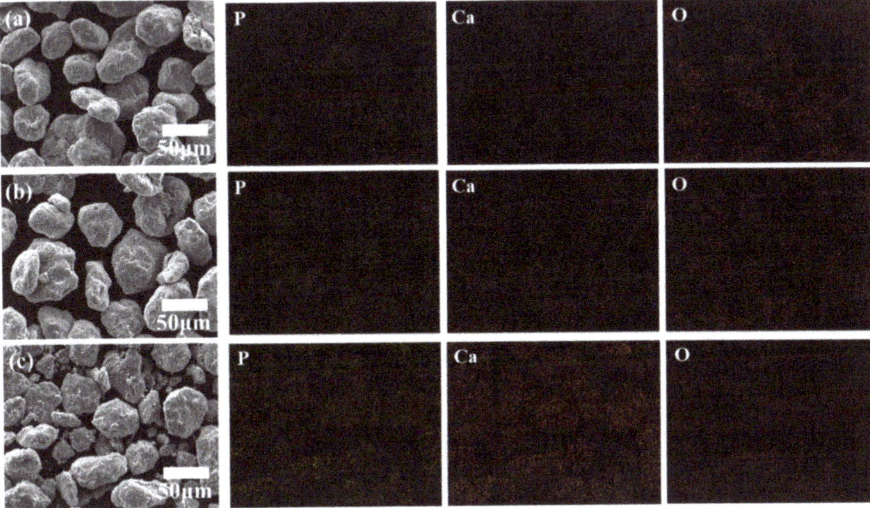

Figure 2. SEM images and EDS analysis of ball-milled Ti6Al4V-EB composites under different weight fraction of the EB; (**a**) T-0.05Ed; (**b**) T-0.5EB and (**c**) T-5EB.

For all cases of Ti6Al4V-EB composites, well-distributed EB and α + β morphology were observed. Regarding the distribution of Al and V, the corresponding concentration ratios for α-Ti and β-Ti were found to be approximately 3:2 and 3:4, respectively (in Figure 3e) [31]. The V concentration in the Ti matrix increased according to the β-Ti phase transformation. The significance in f_β is closely related to elastic modulus, which is an important factor as a biomedical material to attain low elastic modulus, because beta phase has a lower elastic modulus than alpha phase [32–34]. The stress shielding effect can occur according to high elastic modulus; therefore, a decrease in elastic modulus is important for biomedical implants [34–37]. The stress shielding effect is defined as the stress transfer between the bone and implant materials which is not homogeneously caused by a difference in elastic modulus. For instance, when the load is applied, the implant with a high elastic modulus mostly absorbs all of the tensile and compressive stresses and bending moments, which were previously applied to the bone; resultantly, osteoporosis or bone loss can occur [36,38,39].

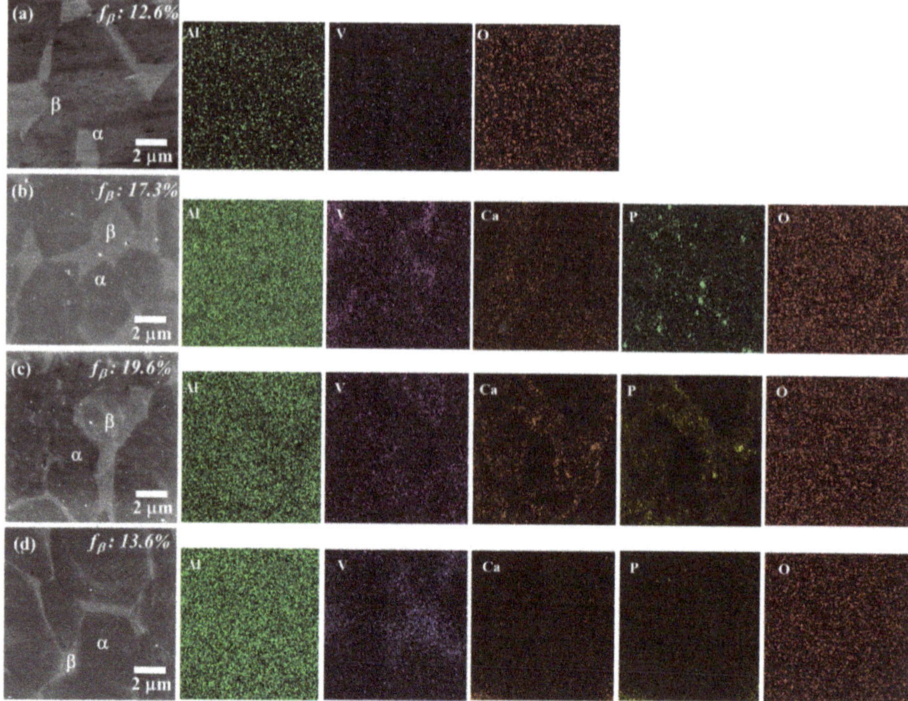

Figure 3. SEM images and EDS analysis of spark plasma sintered (SPSed) Ti6Al4V and Ti6Al4V-EB composites under different weight fraction of the EB; (**a**) Ti6Al4V; (**b**) T-0.05EB; (**c**) T-0.5EB; and (**d**) T-5EB (*f* is a fraction of β-Ti phase).

Figure 4 shows the XRD analysis of the starting materials and Ti6Al4V-EB composite before and after SPS to identify the elements and the phase distribution. The major peaks are attributed to α-Ti (marked by black square), β-Ti (marked by green square), and hydroxyapatite (marked by a red square). All peaks are in good agreement with each standard spectrum from the JCPDS database; α-Ti peaks displayed the (100), (002), (101), (102), and (110) reflection peak around 2θ = 35.4, 38.6, 40.5, 53.4, and 63.7°, respectively (JCPDS Card No. 00-044-1294 and 00-009-0098). Although the peaks of hydroxyapatite for the composite powder and SPSed composites were shrouded in major peaks of Ti due to a small amount of EB, peaks from 2θ = 32° to 2θ = 35.2° for hydroxyapatite (JCPDS Card No. 00-009-0432) were detected, as shown in Figure 3. According to the literature, for sintering up to 1200 °C, hydroxyapatite can prevent decomposition into metastable phases such as tri-calcium phosphate during the sintering [40]. After SPS at 1000 °C, the main peaks were clearly detected for Ti6Al4V-EB composites. Several β-phases within the Ti6Al4V/EB composites, the (110) reflection peak of which is around 2θ = 39.7°, are displayed in Figure 4b. The XRD peak of the β-Ti phase observed in the T-5EB sample subjected to SPS is shifted to a lower angular position, because more V contents in the matrix, with a lower atomic radius of 0.132 nm, were dissolved in the β-Ti phase. After SPS, while 2θ was increased from 40.48 to 40.51°, the peak shifted towards a smaller angle θ of 0.03°, while when 2θ was increased from 38.40 to 38.41°, the peak shifted towards a smaller angle θ of 0.01°.

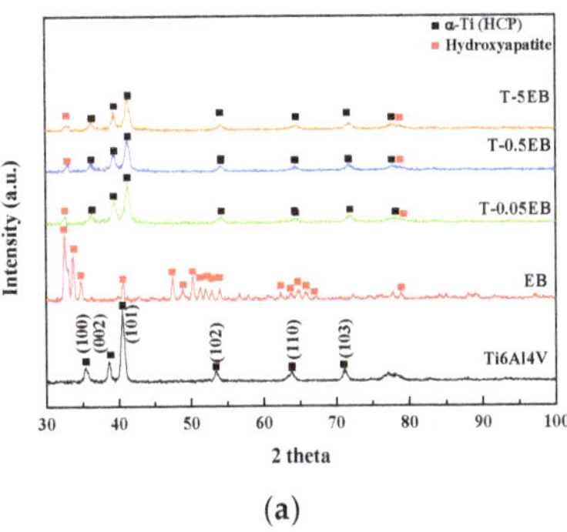

(a)

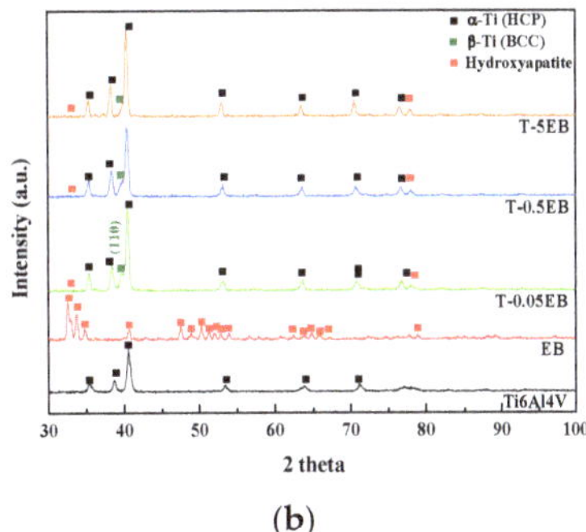

(b)

Figure 4. XRD analysis of the (**a**) starting powders and the Ti6Al4V-EB composite powders, and (**b**) SPSed composites.

3.2. Mechanical Properties

To estimate the mechanical characteristics for biomedical implants, Vickers hardness and elastic modulus of the Ti6Al4V-EB composites, a function of EB contents, are shown in Figure 5a. For accuracy, the mean average values of the elastic modulus were calculated from ultrasonication method and nanoindentation curves. The increase in Vickers hardness by means of EB contents is because the hardness of hydroxyapatite contained in EB (nanohardness of 5–6 GPa and Vickers hardness of 460–480 HV) is higher than that of the Ti matrix (nanohardness of ~2 GPa and Vickers hardness of 350–400 HV) [41–44]. In addition, nanoscale hydroxyapatite was uniformly dispersed in the Ti6Al4V matrix, thereby strengthening particle effects on the hardness increment. The hardness increases as the amount of EB added increases; this increment is based on the previous studies which showed that the hardness can be increased by adding a ceramic material such as EB to the Ti matrix. When hydroxyapatite particles from EB are added to the Ti matrix, the hydroxyapatite particles are allowed to act as a barrier to dislocations [29,45]. Therefore, the values of Vickers hardness for Ti6Al4V-EB composites have been shown to range from 503.3 to ~690.1 HV, which is 1.4–2 times higher than that of pure Ti6Al4V [44]. However, the values of elastic modulus have displayed a different trend compared to those of Vickers hardness. In general, elastic modulus values for Ti6Al4V-EB composites, which range from 133.2 to 139.7 GPa, are slightly higher than those of Ti6Al4V (114 GPa); however, it depends on the fraction of β-Ti among the composites (in Figure 2). Therefore, the composite containing the highest f_β of 19.6% showed the lowest elastic modulus value of 130.5 GPa, rather than the composite containing a f_β of 13.6% with the 139.7 GPa. In addition, oxygen from EB is a typical stabilizing element for α-Ti—that is, as the amount of EB increased, the fraction of α-Ti having a relatively high elastic modulus was increased due to the stabilizing element. Thus, the elastic moduli of the SPSed composites also increased. Further, the elastic moduli of the SPSed samples were affected by the density; the Ti6Al4V-0.5EB composites showed the lowest elastic moduli due to having a relatively low density of 95%. Consequently, this result showed that the proper selection of process conditions for the fabrication of Ti6Al4V matrix composites could have high strength and suitable elastic moduli for use as biomedical implants. For instance, when the composites are sintered above 1200 °C, hydroxyapatite can phase transform to a meta-stable phase such as tricalcium phosphate, which may lead to biodegradation in the human body [40].

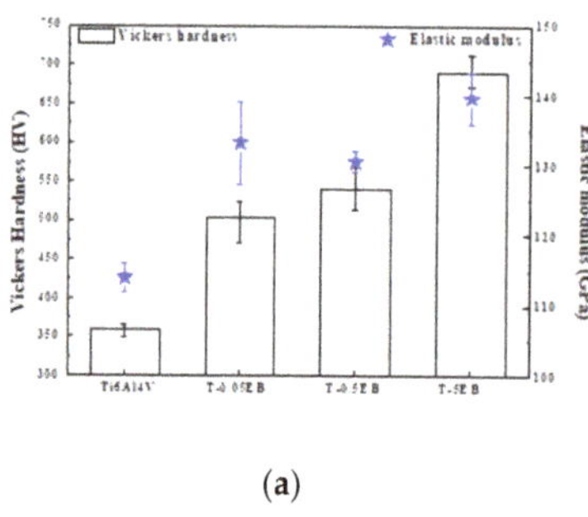

(a)

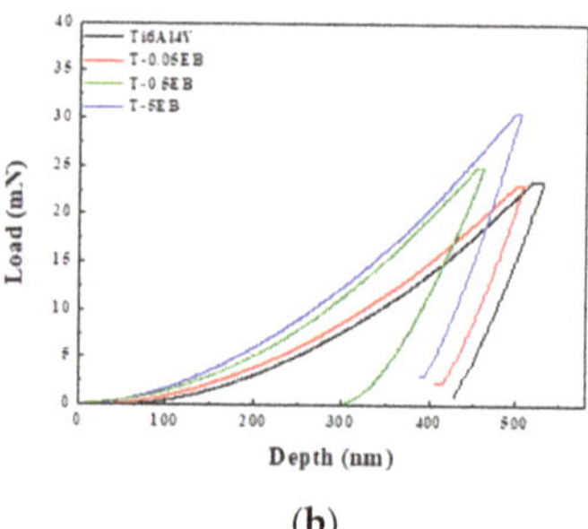

(b)

Figure 5. Mechanical properties of the Ti6Al4V and Ti6Al4V/HB composites. (**a**) Vickers hardness and elastic modulus, and (**b**) stress-strain curves from nanoindentation.

4. Conclusions

In this study, Ti6Al4V powder, which was used as a metal matrix of biocomposites and EB powder with excellent bioactivity, was successfully manufactured using powder metallurgy and SPS. The Ti6Al4V-EB composite powders as a function of EB mass fraction (0.05, 0.5, and 5 wt.%) were ball-milled for 12 h, and the bulk composites were consolidated using SPS at 1000 °C. The microstructures with phase analysis and mechanical properties were analyzed using the sintered Ti6Al4V-EB composites.

(1) The existence of hydroxyapatite, the major component of equine bones, in both Ti6Al4V-EB composite powder and sintered Ti6Al4V-EB composites, was confirmed by SEM-EDS and XRD analysis.

(2) The hardness of the Ti6Al4V-EB composites increased as the EB contents increased owing to uniformly distribution of EB in the Ti6Al4V matrix. The composites SPSed at 1000 °C, which is the beta-phase transformation temperature, provided well-fabricated specimens and showed reasonable mechanical properties.

(3) The composites containing 0.5 wt.% EB exhibited Vickers hardness and elastic modulus of 540.6 HV and 130.5 GPa, which are high strength and reasonable stiffness values for biomedical implants. Slightly high elastic modulus values of the composites can cause stress shielding problems compared to Ti6Al4V (110 GPa).

(4) Ca, P, and O constituting the hydroxyapatite were detected on the surface of all Ti6Al4V-EB composites, which is no change of surface components before and after sintering due to discharge plasma sintering. Therefore, this study can suggest that the Ti6Al4V-EB composites have high bioactivity by increasing the bonding strength between implant and bone.

Author Contributions: Formal analysis, W.J.; Methodology, W.J.; Resources, H.C.; Supervision, S.-E.S.; alidation, H.C.; Writing–original draft, W.J.; Writing–review & editing, S.-E.S. All authors have read and agreed to the published version of the manuscript.

Funding: This research was funded by Korean Ministry of Trade, Industry and Energy (MOTIE), grant number: P0002019.

Acknowledgments: This research was funded and conducted under "the Competency Development Program for Industry Specialists" of the Korean Ministry of Trade, Industry and Energy (MOTIE), operated by Korea Institute for Advancement of Technology (KIAT)(No. P0002019, HRD Program for High Value-Added Metallic Material Expert).

Conflicts of Interest: The authors declare no conflict of interest.

References

1. Wally, Z.J.; Van Grunsven, W.; Claeyssens, F.; Goodall, R.; Reilly, G.C. Porous titanium for dental implant applications. *Metals* **2015**, *5*, 1902–1920. [CrossRef]
2. Yemisci, I.; Mutlu, O.; Gulsoy, N.; Kunal, K.; Atre, S.; Gulsoy, H.O. Experimentation and analysis of powder injection molded Ti10Nb10Zr alloy: A promising candidate for electrochemical and biomedical application. *J. Mater. Res. Technol.* **2019**, *8*, 5233–5245. [CrossRef]
3. Slokar, L.; Štrkalj, A.; Glavaš, Z. Synthesis of Ti-Zr alloy by powder metallurgy. *Eng. Rev.* **2019**, *39*, 115–123. [CrossRef]
4. Wu, S.; Liu, X.; Yeung, K.W.; Guo, H.; Li, P.; Hu, T.; Chung, C.Y.; Chu, P.K. Surface nano-architectures and their effects on the mechanical properties and corrosion behavior of Ti-based orthopedic implants. *Surf. Coat. Technol.* **2013**, *233*, 13–26. [CrossRef]
5. Hsieh, M.-F.; Perng, L.-H.; Chin, T.-S. Hydroxyapatite coating on Ti6Al4V alloy using a sol–gel derived precursor. *Mater. Chem. Phys.* **2002**, *74*, 245–250. [CrossRef]
6. Best, S.M.; Porter, A.E.; Thian, E.S.; Huang, J. Bioceramics: Past, present and for the future. *J. Eur. Ceram. Soc.* **2008**, *28*, 1319–1327. [CrossRef]
7. Sousa, S.R.; Barbosa, M.A. Effect of hydroxyapatite thickness on metal ion release from Ti6Al4V substrates. *Biomaterials* **1996**, *17*, 397–404. [CrossRef]
8. Gu, Y.W.; Khor, K.A.; Cheang, P. In vitro studies of plasma-sprayed hydroxyapatite/Ti-6Al-4V composite coatings in simulated body fluid (SBF). *Biomaterials* **2003**, *24*, 1603–1611. [CrossRef]
9. Linez-Bataillon, P.; Monchau, F.; Bigerelle, M.; Hildebrand, H.F. In vitro MC3T3 osteoblast adhesion with respect to surface roughness of Ti6Al4V substrates. *Biomol. Eng.* **2002**, *19*, 133–141. [CrossRef]
10. Meachim, G.; Williams, D. Changes in nonosseous tissue adjacent to titanium implants. *J. Biomed. Mater. Res.* **1973**, *7*, 555–572. [CrossRef]
11. Fojt, J. Ti–6Al–4V alloy surface modification for medical applications. *Appl. Surf. Sci.* **2012**, *262*, 163–167. [CrossRef]
12. Evans, S.L.; Gregson, P.J. The effect of a plasma-sprayed hydroxyapatite coating on the fatigue properties of Ti-6Al-4V. *Mater. Lett.* **1993**, *16*, 270–274. [CrossRef]
13. Kweh, S.W.K.; Khor, K.A.; Cheang, P. Plasma-sprayed hydroxyapatite (HA) coatings with flame-spheroidized feedstock: Microstructure and mechanical properties. *Biomaterials* **2000**, *21*, 1223–1234. [CrossRef]
14. Wen, C. *Surface coating and modification of metallic biomaterials*; Woodhead Publishing: Cambridge, UK, 2015.
15. Sidane, D.; Chicot, D.; Yala, S.; Ziani, S.; Khireddine, H.; Iost, A.; Decoopman, X. Study of the mechanical behavior and corrosion resistance of hydroxyapatite sol–gel thin coatings on 316 L stainless steel pre-coated with titania film. *Thin Solid Films* **2015**, *593*, 71–80. [CrossRef]
16. Pilliar, R.; Deporter, D.; Watson, P.; Pharoah, M.; Chipman, M.; Valiquette, N.; Carter, S.; De Groot, K. The effect of partial coating with hydroxyapatite on bone remodeling in relation to porous-coated titanium-alloy dental implants in the dog. *J. Dent. Res.* **1991**, *70*, 1338–1345. [CrossRef]
17. Wiria, F.; Leong, K.; Chua, C.; Liu, Y. Poly-ε-caprolactone/hydroxyapatite for tissue engineering scaffold fabrication via selective laser sintering. *Acta. Biomater.* **2007**, *3*, 1–12. [CrossRef]
18. Guizzardi, S.; Montanari, C.; Migliaccio, S.; Strocchi, R.; Solmi, R.; Martini, D.; Ruggeri, A. Qualitative assessment of natural apatite in vitro and in vivo. *J. Biomed. Mater. Res.* **2000**, *53*, 227–234. [CrossRef]
19. Arifin, A.; Sulong, A.B.; Muhamad, N.; Syarif, J.; Ramli, M.I. Material processing of hydroxyapatite and titanium alloy (HA/Ti) composite as implant materials using powder metallurgy: A review. *Mater. Des.* **2014**, *55*, 165–175. [CrossRef]
20. Knabe, C.; Berger, G.; Gildenhaar, R.; Klar, F.; Zreiqat, H. The modulation of osteogenesis in vitro by calcium titanium phosphate coatings. *Biomaterials* **2004**, *25*, 4911–4919. [CrossRef]
21. Li, F.; Jiang, X.; Shao, Z.; Zhu, D.; Luo, Z. Research Progress Regarding Interfacial Characteristics and the Strengthening Mechanisms of Titanium Alloy/Hydroxyapatite Composites. *Materials* **2018**, *11*, 1391. [CrossRef]
22. Hameed, P.; Gopal, V.; Bjorklund, S.; Ganvir, A.; Sen, D.; Markocsan, N.; Manivasagam, G. Axial Suspension Plasma Spraying: An ultimate technique to tailor Ti6Al4V surface with HAp for orthopaedic applications. *Colloids Surf. B* **2019**, *173*, 806–815. [CrossRef] [PubMed]

23. Matsumoto, T.; Kawakami, M.; Kuribayashi, K.; Takenaka, T.; Minamide, A.; Tamaki, T. Effects of sintered bovine bone on cell proliferation, collagen synthesis, and osteoblastic expression in MC3T3-E1 osteoblast-like cells. *J. Orthop. Res. Off. Publ. Orthop. Res. Soc.* **1999**, *17*, 586–592. [CrossRef] [PubMed]
24. Ooi, C.Y.; Hamdi, M.; Ramesh, S. Properties of hydroxyapatite produced by annealing of bovine bone. *Ceram. Int.* **2007**, *33*, 1171–1177. [CrossRef]
25. Jang, K.-J.; Cho, W.J.; Seonwoo, H.; Kim, J.; Lim, K.T.; Chung, P.-H.; Chung, J.H. Development and Characterization of Horse Bone-derived Natural Calcium Phosphate Powders. *J. Biosyst. Eng.* **2014**, *39*, 122–133. [CrossRef]
26. Yan, W.-Q.; Nakamura, T.; Kawanabe, K.; Nishigochi, S.; Oka, M.; Kokubo, T. Apatite layer-coated titanium for use as bone bonding implants. *Biomaterials* **1997**, *18*, 1185–1190. [CrossRef]
27. Belcarz, A.; Bieniaś, J.; Surowska, B.; Ginalska, G. Studies of bacterial adhesion on TiN, SiO2–TiO2 and hydroxyapatite thin layers deposited on titanium and Ti6Al4V alloy for medical applications. *Thin Solid Films* **2010**, *519*, 797–803. [CrossRef]
28. Uezono, M.; Takakuda, K.; Kikuchi, M.; Suzuki, S.; Moriyama, K. Hydroxyapatite/collagen nanocomposite-coated titanium rod for achieving rapid osseointegration onto bone surface. *J. Biomed. Mater. Res. B* **2013**, *101B*, 1031–1038. [CrossRef]
29. Niespodziana, K.; Jurczyk, K.; Jakubowicz, J.; Jurczyk, M. Fabrication and properties of titanium–hydroxyapatite nanocomposites. *Mater. Chem. Phys.* **2010**, *123*, 160–165. [CrossRef]
30. Benjamin, J.; Volin, T. The mechanism of mechanical alloying. *Metall. Trans.* **1974**, *5*, 1929–1934. [CrossRef]
31. Semiatin, S.; Knisley, S.; Fagin, P.; Barker, D.; Zhang, F. Microstructure evolution during alpha-beta heat treatment of Ti-6Al-4V. *Metall. Mater. Trans. A* **2003**, *34*, 2377–2386. [CrossRef]
32. Prendergast, P.; Taylor, D. Stress analysis of the proximo-medial femur after total hip replacement. *J. Biomed. Eng.* **1990**, *12*, 379–382. [CrossRef]
33. Sumner, D.R.; Galante, J.O. Determinants of stress shielding. *Clin. Orthop. Relat. Res.* **1991**, *274*, 202–212. [CrossRef]
34. Van Rietbergen, B.; Huiskes, R.; Weinans, H.; Sumner, D.; Turner, T.; Galante, J. The mechanism of bone remodeling and resorption around press-fitted THA stems. *J. Biomech.* **1993**, *26*, 369–382. [CrossRef]
35. Schreurs, B.W.; Huiskes, R.; Buma, P.; Slooff, T.J.J.H. Biomechanical and histological evaluation of a hydroxyapatite-coated titanium femoral stem fixed with an intramedullary morsellized bone grafting technique: An animal experiment on goats. *Biomaterials* **1996**, *17*, 1177–1186. [CrossRef]
36. Niinomi, M.; Nakai, M. Titanium-based biomaterials for preventing stress shielding between implant devices and bone. *Inter. J. Biomater.* **2011**, *2011*. [CrossRef] [PubMed]
37. Ning, C.; Zhou, Y. On the microstructure of biocomposites sintered from Ti, HA and bioactive glass. *Biomaterials* **2004**, *25*, 3379–3387. [CrossRef] [PubMed]
38. Dujovne, A.; Bobyn, J.; Krygier, J.; Miller, J.; Brooks, C. Mechanical compatibility of noncemented hip prostheses with the human femur. *J. Arthroplasty* **1993**, *8*, 7–22. [CrossRef]
39. Niinomi, M. Mechanical properties of biomedical titanium alloys. *Mater. Sci. Eng. A* **1998**, *243*, 231–236. [CrossRef]
40. Ruys, A.; Wei, M.; Sorrell, C.; Dickson, M.; Brandwood, A.; Milthorpe, B. Sintering effects on the strength of hydroxyapatite. *Biomaterials* **1995**, *16*, 409–415. [CrossRef]
41. Zhao, X.; Liu, Q.; Yang, J.; Zhang, W.; Wang, Y. Sintering Behavior and Mechanical Properties of Mullite Fibers/Hydroxyapatite Ceramic. *Materials* **2018**, *11*, 1859. [CrossRef]
42. Kumari, R.; Scharnweber, T.; Pfleging, W.; Besser, H.; Majumdar, J.D. Laser surface textured titanium alloy (Ti–6Al–4V)–Part II–Studies on bio-compatibility. *Appl. Surf. Sci.* **2015**, *357*, 750–758. [CrossRef]
43. Thijs, L.; Verhaeghe, F.; Craeghs, T.; Humbeeck, J.V.; Kruth, J.-P. A study of the microstructural evolution during selective laser melting of Ti–6Al–4V. *Acta. Mater.* **2010**, *58*, 3303–3312. [CrossRef]
44. Woo, K.D.; Kim, S.H.; Kim, J.Y.; Park, S.H. Fabrication and Biomaterial Characteristics of HA added Ti-Nb-HA Composite Fabricated by Rapid Sintering. *Korean J. Met. Mater.* **2012**, *50*, 86–91. [CrossRef]
45. Bovand, D.; Yousefpour, M.; Rasouli, S.; Bagherifard, S.; Bovand, N.; Tamayol, A. Characterization of Ti-HA composite fabricated by mechanical alloying. *Mater. Des. (1980–2015)* **2015**, *65*, 447–453. [CrossRef]

MDPI
St. Alban-Anlage 66
4052 Basel
Switzerland
Tel. +41 61 683 77 34
Fax +41 61 302 89 18
www.mdpi.com

Metals Editorial Office
E-mail: metals@mdpi.com
www.mdpi.com/journal/metals

www.ingramcontent.com/pod-product-compliance
Lightning Source LLC
LaVergne TN
LVHW070648170726
843515LV00004B/353

* 9 7 8 3 0 3 6 5 4 9 3 5 4 *